T0074676

# Free Agents

# Free Agents

## HOW EVOLUTION GAVE US FREE WILL

KEVIN J. MITCHELL

PRINCETON UNIVERSITY PRESS
PRINCETON & OXFORD

Published by Princeton University Press
41 William Street, Princeton, New Jersey 08540
99 Banbury Road, Oxford OX2 6JX

press.princeton.edu



Library of Congress Cataloging-in-Publication Data

Names: Mitchell, Kevin J. (Professor of genetics), author.
Title: Free agents : how evolution gave us free will / Kevin J. Mitchell.
Description: Princeton : Princeton University Press, [2023] | Includes bibliographical references.
Identifiers: LCCN 2022061071 (print) | LCCN 2022061072 (ebook) | ISBN 9780691226231 (hardback) | ISBN 9780691226224 (ebook)
Subjects: LCSH: Free will and determinism—Physiological aspects. | Brain—Evolution. | Neurosciences. | Evolutionary psychology. | BISAC: SCIENCE / Life Sciences / Neuroscience | COMPUTERS / Artificial Intelligence / General
Classification: LCC BF621 .M583 2023 (print) | LCC BF621 (ebook) | DDC 123/.5—dc23/eng/20230415
LC record available at https://lccn.loc.gov/2022061071
LC ebook record available at https://lccn.loc.gov/2022061072

British Library Cataloging-in-Publication Data is available

Editorial: Alison Kalett and Hallie Schaeffer
Jacket: Karl Spurzem
Production: Jacqueline Poirier
Publicity: Maria Whelan and Kate Farquhar-Thomson
Copyeditor: Gail Naron Chalew

This book has been composed in Arno

Printed on acid-free paper. ∞

Printed in the United States of America

10 9 8 7 6 5 4 3 2 1

For Adam and Ethan, with all my love

# CONTENTS

# PREFACE

The question of whether humans have free will has been chewed over by philosophers and theologians for millennia without any clear conclusion or even any agreed-on articulation of the question. Scientists have typically shied away from it to focus on more tractable problems, but recent advances and discoveries, in neuroscience in particular, are changing that.

We are learning more and more details of how patterns of brain activity control behavior, how animals—including humans—make decisions, and how neural circuits accumulate evidence, weigh alternatives, signal value and salience and confidence, and instigate actions. But as that decision-making machinery is being revealed, it seems harder and harder to escape the paradoxical conclusion that we really are *just* machines. Decisions are clearly being made, but where do "we" fit into the picture?

Indeed, it is fashionable among neuroscientists to declare, "Free will is an illusion." Not only do we not have it but also there is no way that we could. We can watch the machine running; we can get in there and tweak it in animals and even in humans to drive decisions and actions. What space does this leave for the mind? There seems to be no need and no room for mental causation of any kind. Who cares what the neural patterns mean when the ions are going to flow where they're going to flow?

Philosophers like Daniel Dennett defend the contrary view that organisms do, in fact, do things for reasons, which are instantiated in the configurations of their neural or biochemical circuitry. However, Dennett maintains that determinism still holds, that no real choice exists, that you could not, at any given moment, "have done otherwise." This compatibilist view of free will is widespread among philosophers and scientists, but it cannot explain how free will or indeed any form of real agency can possibly exist in a deterministic universe. Instead, compatibilists argue that, even though humans actually have no choice, we can treat people as if they do because what goes on in their brains is so complicated as to be completely unpredictable in practice. We can mount a good-enough simulation of free will to form the basis for our system of moral responsibility, even if it is fundamentally illusory.

I do not think that either the strictly determinist *or* the compatibilist position is satisfactory. Both say free will is an illusion to some extent. At an even deeper level, physicists like Brian Greene and Stephen Hawking assert that the workings of our brains are as determined as the orbits of the planets. The laws of physics dictate the interactions of all the particles in our brains, as in any other piece of matter. Even if real randomness exists, it is merely one more physical (though uncaused) cause playing out in the system. Such a view is the ultimate expression of reductionism, the philosophy that has tacitly held sway in biology for centuries.

The implication of reductionism is that biology is not really a science unto itself but is just complicated chemistry, and chemistry is just complicated physics (and psychology is just complicated neuroscience). Causation only flows up, from the interactions at the lowest levels of the system. If we keep digging deeper and deeper we will reveal the true workings of life: indeed, we will not only explain life—we will explain it away. We will show that living things differ from nonliving things only in being more complicated.

For me, that idea is not just wrong—it's wrong-headed. A purely reductionist, mechanistic approach to life completely misses the point. On the contrary: basic laws of physics that deal only with energy and matter and fundamental forces *cannot explain what life is* or its defining property: living organisms do things, for reasons, as causal agents in their own right. They are driven not by energy but by information. And the meaning of that information is embodied in the structure of the system itself, based on its history. In short, there are fundamentally distinct types of causation at play in living organisms by virtue of their organization.

My goal in this book is to explore how living things come to have this ability to choose, to autonomously control their own behavior, to act as causes in the world. The key to this effort, in my view, is to take an evolutionary perspective. If we want to understand how choice and free will could exist in a physical universe, let's look at the details of how they actually came to exist. The book therefore tracks how agency evolved—from the origin of life itself, through the invention of nervous systems, the subsequent elaboration of decision-making and action selection systems, and the eventual emergence of the kind of conscious cognitive control in humans that we refer to as "free will."

Along the way, I address the metaphysical issues that supposedly present fundamental barriers to the possibility of agency. My aim is to show that, in thinking about these issues, we are not limited either to a simplistic physical determinism, in which all causes are located at the level of atoms or quantum fields, or to some kind of magical dualism, where we have to invoke immaterial forces to rescue our own agency.

Instead, I present a conceptual framework that aims to naturalize the concept of agency by grounding the otherwise vague or even mystical-sounding concepts of purpose, meaning, and value. The truth is that, far from being unscientific, those concepts are crucial to understanding what life is, how true agency can exist, and what sorts of freedoms or limitations we actually have as human beings.

This book surveys findings and thinking from many fields of science and philosophy in a way intended to be accessible to nonspecialists. It is not meant to be an exhaustive intellectual history but instead an overview of things as I see them—it is really an extended argument for a way of thinking about the issue of agency. It builds on and attempts to synthesize the work and thinking of countless others, only a tiny fraction of whom are mentioned by name. The bibliography at the end of the book provides pointers to further reading, rather than a comprehensive list of the vast body of relevant literature. I offer my apologies to authors whose work is not cited.

The framing I present has important implications for how we think of who we are as human beings, how we understand our own decision-making processes, and how we recognize the many ways in which our individual agency can be enhanced or infringed. It is also relevant to more fundamental issues in biology: what it means to be a living organism, striving to persist in a hostile universe; how cognition is structured to support adaptive behavior; and how causal knowledge grants causal power to act in the world. In short, in what follows, I hope to show that the story of agency is the story of life itself.

# ACKNOWLEDGMENTS

I am deeply indebted to many friends, relatives, and colleagues for extremely helpful discussions and feedback on all or parts of earlier versions of this book; they include Philip Ball, Melanie Challenger, Keith Farnsworth, Alicia Juarrero, Gary Marcus, James Marshall, Anne Sophie Meincke, Lynn Mitchell, Melanie Mitchell (no relation!), Sean Mitchell, Thomas Mitchell, Henry Potter, David O'Regan, Siobhan Roche, Liberty Severs, Lee Smolin, Clelia Verde, and Tony Zador. Special thanks to all the members of the Representations group—including Francis Fallon, Celeste Kidd, John Krakauer, Tomás Ryan, Mark Sprevak, and Becky Wheeler—for our always stimulating discussions that have informed this work. And to all members of the Basal Cognition group for the same, especially Fred Keijzer, Matthew Sims, and Caroline Stankozi. I am very grateful for the many discussions with an amazing community on Twitter that have helped shape my thinking on diverse topics; this community includes Bjorn Brembs, Matteo Carandini, Sandeep Robert Datta, Erik Hoel, Yogi Jaeger, Yohan John, Luiz Pessoa, Maxwell Ramstead, Adam Safron, and many others. Special thanks to my agent Will Francis and my editor Alison Kalett for all their help in bringing this book to print. And, finally, to Siobhan Roche, for putting up with my philosophical ruminations on endless lockdown walks.

# Free Agents

# 1

# Player One

Are we the authors of our own stories? Or is our apparent freedom of choice really an illusion? These questions were brought home to me recently as I was watching my son play a video game—one where you wander around an open world, meeting interesting denizens of one type or another (and killing quite a few of them). As I watched, his character entered a tavern and approached the bartender, who offered a generic greeting. The game then threw up some options for things you could say in reply to get information about the prospects for fortune and glory in the surrounding territory.

In this exchange, my son's possibilities for action were limited by the game, but he *was* really making choices among them, and these choices then affected how the conversation went and what would subsequently unfold. His decisions were based on his overall goal in the game, the tension between his goals of taking some immediate action or to keep exploring, his need to have enough information to make a decision with confidence, the risk of biting off more than he could chew and losing his hard-won stuff: all these considerations fed into the decisions he made. He had his reasons and he acted on them, just like you or I do every day, all day long.

The bartender, in contrast, was not making choices. He was a classic "non-player character," an NPC. His responses were completely determined by his programming: he had no degrees of freedom. His actions were merely the inevitable outcome of a flow of electrons through the circuits of the game console, constrained by the rules encoded in the software. Even the more sophisticated NPCs in the game, including the monster that eventually caramelized my son's avatar, were similarly constrained. The monster's actions—even in the fast-moving melee—were determined by the software programming and mediated by the electronic components in the console.

Thus the NPCs only *appear* to be making choices. They're not autonomous entities like us: they're just a manifestation of lots of lines of code, implemented in the physical structure of the computer chips. Their behavior is entirely determined by the inputs they get and their preprogrammed responses. We, in contrast, are causes of things in our own right. We have *agency*: we make our own choices and are in charge of our own actions.

At least it seems that way. It certainly feels like we have "free will," like we make choices, like we are in control of our actions. That's pretty much what we do all day—go around making decisions about what to do. Some are trivial, like what to have for breakfast; some are more meaningful, like what to say or do in social or professional situations; and some are momentous, like whether to accept a job offer or a marriage proposal. Some we deliberate on consciously, and others we perform on autopilot—but *we* still perform them. Of course, our options may be more or less constrained (or informed) by all kinds of factors at any given moment, but generally we feel like the authors of our own actions.

And we interpret other people's behavior in terms of their reasons for selecting different actions—their intentions, beliefs, and desires that make up the content of their mental states. We constantly analyze each other's motives and habits and character, looking for explanations and predictors of their behavior and the decisions they make. Why people act the way they do is ultimately the theme of most entertainment, from Dostoyevsky to *Big Brother*. All this rests on the view that we are not just acted on—we are actors. Things don't just happen to us, in the way they happen to rocks or spoons or electrons: *we do things*.

The problem is that, if you think about this view for too long, it becomes difficult to escape a discomfiting thought. After all, like the NPCs, our decisions, however complex they may be, are mediated by the flow of electrical ions through the circuits of our brains and thus are constrained by our own "programming," by how our circuits are configured.

Unless you invoke an immaterial soul or some other ethereal substance or force that is really in charge—call it spirit or simply mind, if you prefer—you cannot escape the fact that our consciousness and our behavior emerge from the purely physical workings of the brain.

There is no shortage of evidence for this from our own experience. If you've ever been drunk, for example, or even just a little tipsy, you've experienced how altering the physical workings of your brain alters your choices and the way you behave. There is a whole industry of recreational drugs—from caffeine to crystal meth—that people take because of the way that physically tweaking the brain's machinery in various ways makes them feel and act. The ultimate consequence in some cases is addiction—perhaps the starkest example of how our actions can sometimes be out of our control.

And, of course, if the machinery of your brain gets physically damaged—as occurs with head injuries, strokes, brain tumors, neurodegenerative disorders, or a host of other kinds of insults—or its function is impaired in other ways, as in conditions such as schizophrenia, depression, or mania, then your ability to choose your actions may also be impaired. In some situations the integrity of your very *self* may be compromised.

We all like to think that we are Player One in this game of life, but perhaps we are just incredibly sophisticated NPCs. Our programming may be complex and subtle enough to make it *seem* as if we are really making decisions and choosing our own actions, but maybe we're just fooling ourselves. Perhaps "we" are just the manifestations of genetic and neural codes, implemented in biological rather than computer hardware. Perhaps we are the victims of a cruel joke, tragic figures in the grip of the Fates. As Gnarls Barkley sang, "Who do you, who do you, who do you think you are? Ha ha ha, bless your soul, you really think you're in control."

## Robots with Personality

In my 2018 book *Innate* I described how we all come pre-wired with a set of innate psychological predispositions. At the most basic level, we all share the profile of human nature. Evolution has shaped the behavior of our species just as much as that of any other. Human nature is encoded in our DNA in a genetic program that specifies the building and wiring of our human brains.

However, the details of that genetic program inevitably vary among individuals. I use the word "inevitably" because there is no way that this variation could not exist. Every time DNA is copied in a cell, including when sperm or egg cells are made, some small number of copying errors or mutations arise. New variations in the DNA sequence thus enter the gene pool in every generation, and—if their effects are tolerated—they can spread through the population over time, leading to the accumulation of genetic variation that we observe.

This leads to the differences that we observe in people's physical traits, such as how tall they are or the shape of their faces or various aspects of their physiology. This variation occurs just the same in the physical structure of their brains and the way they function. The fact that all these traits are affected by genetic variation explains why people who are related to each other resemble each other more than do unrelated people, both physically and psychologically. So, even though the "canonical" human genome (which doesn't really exist anywhere) encodes a program to build a canonical human brain, your particular genome encodes a program to build *a brain like yours*.

But not exactly like yours. The program in your genome does not encode one particular outcome, specified down to the level of individual nerve cells or synaptic connections between them. It does not encode the outcome at all, in fact: it just encodes a set of biochemical rules that, when played out over the complicated processes of development, will tend to result in an outcome within a certain range. Exactly how these processes played out in your specific case was also affected by all kinds of random events during development that added considerable variation. If you ran the program again, you would not get exactly the same outcome. Even the brains of identical twins who share the same genetic program are quite distinct from each other already at birth.

All this means that the way your brain is wired is affected by millions of years of common evolution, by the specific genetic variations that you carry, and by the unique trajectory of development that occurred

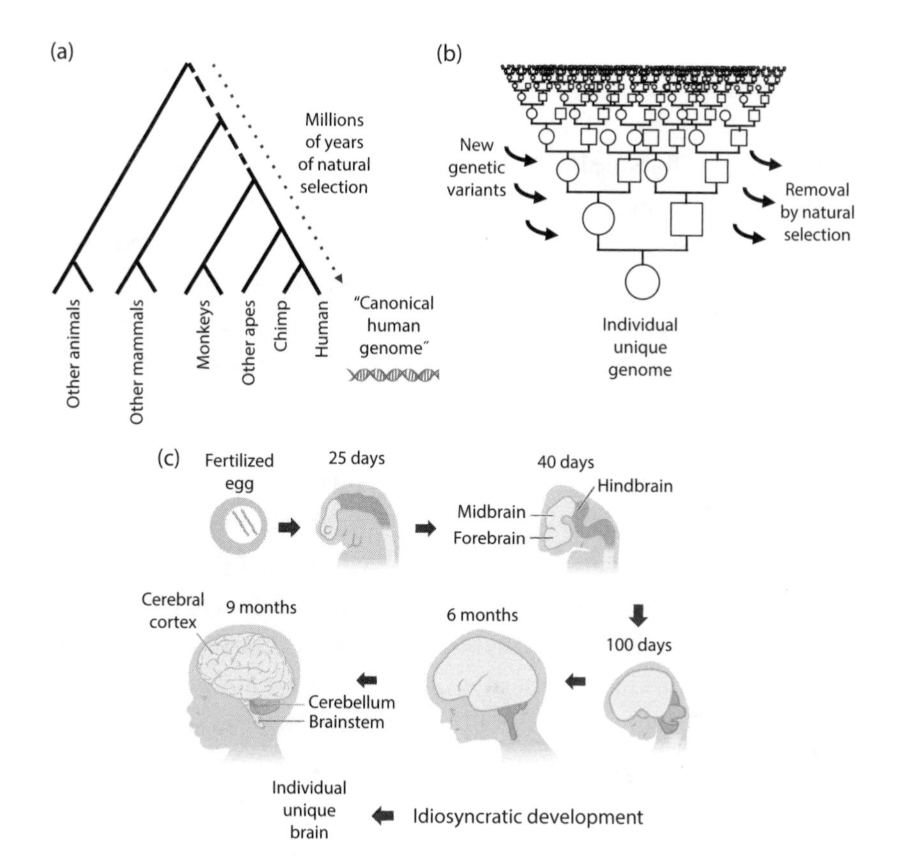


FIGURE 1.1. The making of you. **(a)** Like all species, the genome of Homo sapiens has been shaped by millions of years of evolution, selecting for all the traits that comprise "human nature" generally. **(b)** Your individual genome is a unique version of the canonical human genome, reflecting the processes of mutation and selection in your specific ancestors. **(c)** The outcome of brain development in any individual is idiosyncratic, shaped by genetic variation and the unique trajectories of development itself. Our individual natures (or innate predispositions) are thus variations on the theme of human nature generally.

while your brain was developing (see Figure 1.1). And the way your brain is wired affects how it works and how you will tend to behave.

We can think of this variation like that in the internal tuning of the behavioral controls of a robot. Imagine you and I were asked to build an autonomous robot that has to make its way in the world—finding fuel, avoiding threats, interpreting sensory information, assessing situations, and deciding among possible actions.

There are all kinds of things we would have to build in our robot for it to accomplish its tasks. It would need some sensors, of course, to detect things in the environment, and it would require motors so it could move around and perform various actions. It would need to be programmed to move toward fuel and away from threats, but it might also require some fancy circuitry for it to recognize which is which. And what would happen if fuel supplies and something threatening happen to be in the same place? It would have to weigh the opportunity versus the risk and make a decision accordingly about where to move. And it would be good if that decision were informed by how much fuel it had left at the time. So some way to monitor its internal states and use them to inform decisions would certainly be beneficial.

A very fancy robot might also be able to learn from experience; for example, it might learn that there tends to be fuel in some particular spot or that some kind of otherwise innocuous stimulus (a rustle in the robot grass perhaps) signals a hidden threat. Now imagine we give our robot another goal: not just to survive but also to find robot love and reproduce. Then it would have to balance the short-term goal of ensuring it has enough fuel with the longer-term goal of finding a mate, all while not getting destroyed by a bigger robot.

All those functions—some means for inferring what is out in the world from the data gathered by its sensors; integrating both external and internal multiple signals to derive a picture of the whole situation; comparing that with the data in its memory bank to help inform its next action; weighing threats versus opportunities, short-term versus long-term goals, and good versus bad outcomes; and eventually picking one action to perform while inhibiting all other possibilities—would have to be configured into its circuitry.

With so many circuits and parameters that could vary, it is inevitable that the way you would tune your robot would differ from how I would tune mine. You might set the threat sensitivity a little higher and the reward sensitivity a little lower. I might tune the circuits in my robot with a different balance between short- and long-term goals. All these settings would manifest as different *patterns of behavior* over time and across contexts. Your robot might appear more cautious than mine. Mine might show more perseverance: it might be willing to work longer for a delayed reward. The robots could differ in how much evidence they need to make a decision (impulsivity), how much they value mating opportunities (sex drive), and even how salient they find novel objects or situations (curiosity). In short, our robots would have personalities, just like you and I do.

And just like you or me, they would not have had any hand in choosing those traits. Even if the robots learn over their lifetime and adapt to the various scenarios they encounter, all this learning is also physically embodied in the configuration of their circuits at the moment they are faced with a decision. The sense of fatalism that this realization engenders is aptly summarized by prominent free will skeptic Sam Harris:

> Take a moment to think about the context in which your next decision will occur: You did not pick your parents or the time and place of your birth. You didn't choose your gender or most of your life experiences. You had no control whatsoever over your genome or the development of your brain. And now your brain is making choices on the basis of preferences and beliefs that have been hammered into it over a lifetime—by your genes, your physical development since the moment you were conceived, and the interactions you have had with other people, events, and ideas. Where is the freedom in this? Yes, you are free to do what you want even now. But where did your desires come from?[1]

The essence of the problem was captured by the famously pessimistic (or some might say realistic) philosopher Arthur Schopenhauer, who said, "A man can do what he will, but not will as he will."[2] Even if we are making choices right now, those choices are not free from all kinds of prior causes or influences, over which we had no control.

1. Sam Harris, *Free will* (New York: Free Press, 2012), 44.
2. Arthur Schopenhauer, *Essay on the freedom of the will* (New York: Dover, 1960), 6.

## The Machine

As a neuroscientist, this kind of existential worry is an occupational hazard. But it gets worse. The more we learn about the mechanisms of perception and cognition and, in particular, of decision making and action selection, the more *mechanistic* it all seems and the less there seems to be for *the mind* to do. How can we even think that *we* are making choices at all, when we can see that the process is the result of just a bunch of gears turning in the machine? What reason is there to think that an entity is in charge?

And, thanks to modern technology, we can actually see the figurative gears turning. Using a variety of neuroimaging tools in humans and animals to track the activity of different neural circuits or brain areas, it is possible to tease out the types of information they carry and the cognitive operations they perform as the organisms or individuals make decisions or select actions. We can, for example, distinguish patterns of neural activity that correlate with (and seem to internally "represent") the accumulation of evidence about something in the world, the degree of certainty attached to some signal, the confidence level in a belief, the adoption of a new goal, the rewards associated with a positive outcome, the learning that happens in response to such rewards, the emotional signals that accompany decision making, the gradual formation of habits, the real-time switch from habitual to goal-directed or exploratory behavior as circumstances change, and on and on. We can see the thinking happening.

We can even, in some circumstances, predict an incipient action before the individual performs it. There are many experimental setups using rodents or monkeys where researchers can track patterns of brain activity, observe a threshold being approached that will result in an action, and even predict (not with complete accuracy but significantly better than chance) what action it will be—whether a rat will turn left or right in a maze, for example.

In humans there is a famous example where an action was not only predicted ahead of time but also before the subject even became consciously aware of having chosen to do it. In these experiments, performed by Benjamin Libet and colleagues in the 1980s, subjects had to randomly decide to move their fingers while watching a clock and while their brainwaves were being recorded by an electroencephalograph. The striking result: the onset of brain activity leading to a movement preceded the reported timing of the conscious awareness of the intention to act by several hundred milliseconds.

Although not relevant to truly deliberative decisions, these findings can still shake your faith in your conscious mind really being in control of your actions. Is the rest of the brain just flattering us, making us feel that we're in charge, like a wily civil servant expertly managing his elected boss?

If pulling back the curtain to expose the neural machinery of decision making at work were not enough of a threat to our egos (in both senses of that word), it is also possible to intervene in the machine—to drive patterns of neural activity from the outside—and *cause* the individual to behave in certain ways.

Famous experiments carried out in the 1940s by neurologist Wilder Penfield and his colleagues in human subjects undergoing brain surgery (who were awake and aware throughout the procedure) showed that stimulating different parts of the cerebral cortex with electrodes could produce all kinds of sensations, emotions, urges, memories, or movements of various parts of the body (see Figure 1.2). This work contributed greatly to the mapping of functions across the brain and reinforced the view of a complex electrical machine *producing* the contents of the mind, rather than being controlled by that mental content.

Similar experiments are possible in animals, but, as in humans, they're a bit crude. Just poking an electrode into a part of the brain and zapping it activates all the neurons in that area in a nonspecific fashion. The brain then attempts to make sense of that mini-explosion of activation, but this process is very different from how neural signaling normally happens. Indeed, within any little chunk of brain, there are hundreds of different types of nerve cells connected in intricate microcircuits designed to carry out diverse sorts of computations. Just blasting them all at once is thus not hugely informative about how these computations mediate cognitive operations.

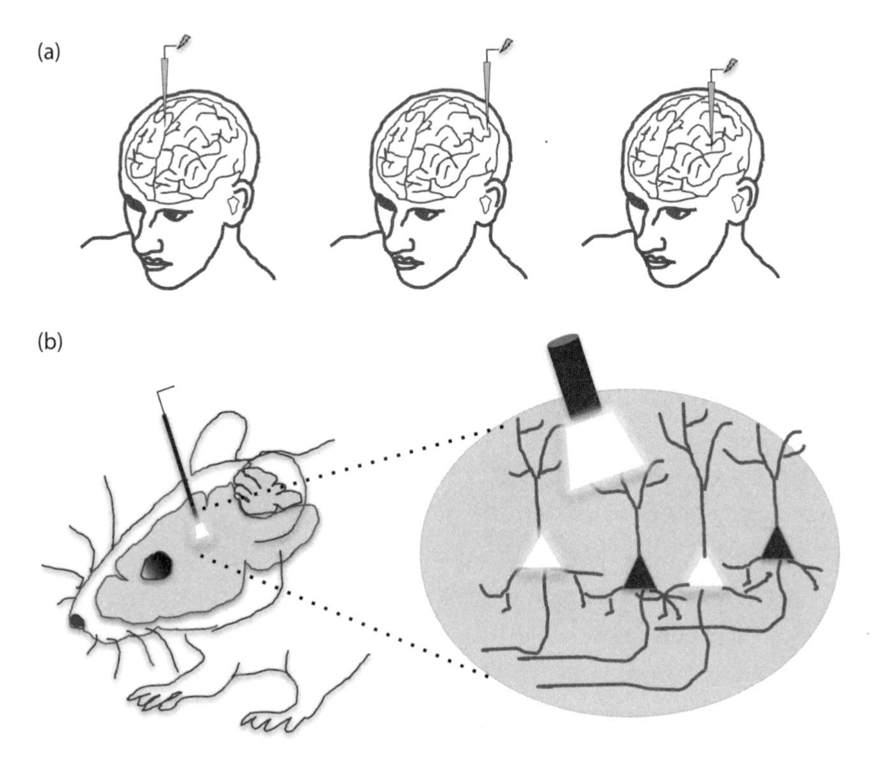


FIGURE 1.2. Brain stimulation. (**a**) Direct stimulation of brain areas in awake subjects can lead to involuntary movements (motor cortex, left), sensory percepts (visual cortex, middle), or even activation of memories (temporal lobe, right), depending on the area stimulated. (**b**) Optogenetic techniques in rodents allow much more specific activation of genetically targeted subsets of neurons, providing a powerful platform to dissect the functions of neural circuits in awake, behaving animals.

Seventy years after Penfield's experiments, the study of the neural systems that control behavior in animals was revolutionized by the invention of molecular tools that allow researchers to drive the activity of very specific subsets of neurons in an animal's brain while observing its behavior in real time. Like most techniques in molecular biology, this one—called optogenetics—borrows from nature. It uses a protein made by blue-green algae that sits in the membrane of the cell and responds to light by opening up a channel through which electrically charged atoms (or ions) can pass. That protein is related to ones that we use in our eyes to detect light, and it is exactly the opening of ion channels in the membrane that drives nerve cells to "fire" or send a sharp electrical signal.

Researchers including Karl Deisseroth, Edward Boyden, and others realized that if they cloned the algal gene that encodes this protein and transferred it to mammalian neurons, they could effectively turn "on" the neurons with exquisite temporal precision by shining a blue light on them. Hooking the piece of DNA that codes for this light-responsive channel protein (called channel-Rhodopsin) to the DNA codes that regulate the expression of all kinds of different genes in the mouse brain enabled them to generate lines of transgenic mice expressing channel-Rhodopsin in extremely specific subsets of neurons in different brain regions.

Shining a light on the relevant bit of the brain—accomplished by threading a minute fiber optic cable through the skull—allows researchers to activate just that specific subset of neurons within the circuit and study the effects on behavior of the animal. Using this technique, specific sets of neurons were identified that, when activated, drive all kinds of behaviors—from general locomotion to more subtle motor actions like reaching or grasping, from aggression to mating, from freezing in fright to lunging attacks on prey that are not present, from eating to sleeping to looking after pups, and on and on.

But this research reaches far beyond directly activating particular actions from the animal's repertoire of behavior. It has made it possible to dissect the cognitive machinery involved in choosing among actions, weighing options, signaling rewards and punishments, judging the reliability of sensory information, assigning a level of certainty or confidence to a decision, using past memories to guide current actions, and selecting one option while inhibiting every other possibility. It is even possible, as my colleague Tomás Ryan and others have done, to implant false memories in an animal's brain that will influence its future behavior. This is not just remote control of what the animal is doing: it is control of what the animal *is thinking*.

It's hard not to look at this growing body of work and see only the machine at work. Driving this circuit or that one either directly causes an action or influences the cognitive operations that the animal—mouse or human or anything else—uses to decide between actions. If we were dissecting a robot in this way we would apply engineering approaches to understand the kinds of information being processed, the control mechanisms configured into the different circuits, and the computations that lead to one output or another. There does not seem to be any need for something like *a mind* in that discussion. There is no real need for *life*, for that matter.

If the circuits just work on physical principles, then who cares what the patterns of activity *mean*? Why does it matter what the mental content associated with a particular pattern of neural activity is, if it is solely the physical configuration of the circuitry that is going to determine what happens next? We may have set out, as neuroscientists, to explain how the workings of the brain generate or realize psychological phenomena, but we are in danger of explaining those phenomena away.

## It's All Just Physics at the End of the Day

If the neuroscientists have it bad, pity the poor physicists, whose existential angst must run much deeper. Where neuroscientists can at least hold onto the view that the circuits in the brain are doing things (whether "*you*" are or not), some physicists claim that even that functionality is an illusion. After all, the brain is made of molecules and atoms that must obey the laws of physics, just like the molecules and atoms in any other bit of matter.

These small bits of matter are pushed and pulled by all the forces acting on them—gravity, electromagnetism, the so-called strong and weak nuclear forces that hold atoms together—and where each atom goes is fully determined by the way those interactions play out. These processes are no doubt complicated, as they would be in any system with so many atoms simultaneously acting on each other, and in practice how the system will evolve is unpredictable—but it is still all driven by the physics. Even at the lower levels of subatomic particles, how the system evolves is captured by the equations of quantum mechanics in a way that many would argue theoretically leaves no room for any other causes to be at play.

So, then, what does it matter what you are thinking? You cannot push the atoms in your brain around with a thought. You cannot override the fundamental laws of physics or exert some ghostly control over the basic constituents of matter. According to this view, the very idea of mental causation—of the content of your thoughts and beliefs and desires mattering in some way—is a naive superstition, a conceptual hangover inherited from philosophers like the famous dualist Rene Descartes.

Here is the late Stephen Hawking on the subject: "Biological processes are governed by the laws of physics and chemistry and therefore are as determined as the orbits of the planets. Recent experiments in neuroscience support the view that it is our physical brain, following the known laws of science, that determines our actions and not some agency that exists outside those laws . . . so it seems that we are no more than biological machines and that free will is just an illusion."[3] Brian Greene, another well-known physicist and author, agrees: "Free will is the sensation of making a choice. The sensation is real, but the choice seems illusory. Laws of physics determine the future."[4]

There are two main flavors of this kind of physical determinism. In the first, the low-level laws of physics rule completely: every aspect of the way the universe and everything in it evolves is fully determined by how these interactions play out. There is no room for any other force and, in particular, no role for any kind of randomness or indeterminacy. This model can be summed up as follows:

current state + laws of physics → next state

The consequences of this view are stark. If you keep on working through from one state to the next, you quickly realize that the current state predicts not just one step ahead but also two, or three, or actually an infinite number. And you can work backward just as easily as forward. If this is really the whole picture, then the entire history of the

3. Stephen Hawking and Leonard Mlodinow, *The grand design* (New York: Bantam Books, 2010), 32.

4. Brian Greene (@bgreene) on Twitter, June 5, 2013, https://twitter.com/bgreene/status/342376183519916033?lang=en.

universe up until now and for the rest of time was *pre*determined from shortly after the Big Bang. Indeed, our conception of time as having a direction goes out the window. The whole universe, over all time, is simply given, as a block: there is no real difference between the past and the future. There are no possibilities—only what has happened and what will happen. This view is known as *hard determinism*.

The implied softer version differs in allowing some randomness or indeterminacy to exist. It holds that the future is *not* fully predetermined by the current state (and certainly not by the initial state of the universe). Here, the past and the future are very different: the past is fixed while the future is a branching web of possibilities, only one line of which will be realized at any choice point.

However, even though the branch that is taken is not predetermined in this model, it is still decided by the low-level interactions of all the atoms and molecules. It is just that some of those interactions are a bit random. You might sum up this view like this:

current state + laws of physics + randomness → next state

The debate over whether there really is any true randomness in physical events has been raging since the days of Einstein and Bohr. When you get down to the quantum level of subatomic particles, weird things happen, and even though the weirdness can be fully accounted for in the equations that physicists use, allowing them to make exquisitely precise predictions, there is no consensus at all about what these equations imply about the fundamental nature of reality.

We'll return to this topic of randomness later. For now, what are the implications of this softer version of determinism? It is often summed up with the pithy line: "every event has a cause." This doesn't seem to align with the idea of random events happening, which would seem not to have a cause, by definition. What this statement really seems to imply is that everything that happens—at a system level—is caused by the interactions of particles at the lowest level, even if some randomness is at play there.

Yet, that view seems to be just as problematic for the idea of organisms like us being in charge of anything that happens. The future may not be written, but if what happens is still decided by how the physical forces play out at the minutest scale of matter, there doesn't seem to be much scope for us to be in control. Even neuroscientist Patrick Haggard, a leader in the study of volition, agrees: "As a neuroscientist, you've got to be a determinist. There are physical laws, which the electrical and chemical events in the brain obey. Under identical circumstances, you couldn't have done otherwise; there's no 'I' which can say 'I want to do otherwise.'"[5]

In hard determinism, there are no causes. The universe just inexorably unfolds according to the laws of physics. If nothing could ever be or have been different, then you cannot point to one thing being a certain way and say it caused something else. The concept just doesn't apply. In *soft determinism*, there are causes—some things could be different, depending on how that little bit of randomness plays out—but all the causes are located at the lowest levels. That lowest level is deemed to be the bedrock of reality.

Some physicists, like Sean Carroll or Sabine Hossenfelder, may be magnanimous enough to allow that descriptions at higher levels of organization are "useful ways of talking about" complicated systems. We can productively do chemistry or biology or psychology with theories and methods that remain at those higher levels. But Carroll maintains that the real truth—the whole truth—resides at the lowest level, with the fundamental physical interactions of the smallest particles. If you had a complete accounting of what is going on down there, then you would not need any other information to fully predict what the system will do: everything happening at the higher levels simply derives or emerges from the low-level dynamics. Every other description is just a kind of coarse-grained picture, a *simplification* or statistical averaging that allows our puny minds to grasp how various systems—like cells or brains or minds—behave, despite all the underlying complexity.

5. Patrick Haggard, Neuroscience, free will and determinism: "I'm just a machine," interview by Tom Chivers, *The Telegraph*, October 12, 2010, https://www.telegraph.co.uk/news/science/8058541/Neuroscience-free-will-and-determinism-Im-just-a-machine.html.

Given the phenomenal successes of modern physics in confirming the predictions of quantum mechanics with eye-watering precision, it is not surprising that the focus has been on continuing to develop and test such theories while not worrying too much about what they mean for the nature of reality. The admonition to "Shut up and calculate!" by quantum physicist David Mermin is effectively the motto of the field. Let the philosophers worry about what it all means, especially for metaphysical concepts like free will.

## The Blame Game

Philosophers, for their part, have been debating the implications of deterministic theories of the physical universe for free will for thousands of years, at least as long ago as Democritus and Epicurus in ancient Greece. That these debates continue today with unabated fervor tells you that they have not yet resolved the issue.

In fairness, free will is a uniquely vexing problem. The phenomenon we are trying to explain—our own experience of having the power to make choices—seems inherently at odds with what we know about how everything else works in the universe. The scientific rejection of the idea of an immaterial soul or spirit that is somehow pulling the strings has left us scrambling to explain instead how the machine could pull its own strings. And the progress of physics into the wonderful weirdness of the quantum realm has only deepened the mysteries of what the machine and the world around it are made of in the first place.

But if philosophy can be excused for not having provided an answer, one might at least have hoped for some consensus on what is the right question. The popular framing, "Do we have free will?" is undermined in an obvious way by a lack of agreed-on definitions. If you define the capacity of free will as being able to make decisions in a way that is necessarily *free from every prior cause*, then you have set an unattainable standard, one that could only be met by supernatural means. Alternatively, if your criterion is merely that a person is doing things *based on causes internal to his or her physical self*, then you have not met the challenge of physical determinism but merely sidestepped it with appeals to complexity and unpredictability.

Less obviously, the question "Do we have free will?" is more deeply undermined by a lack of clarity of the terms "we" and "have." We cannot profitably approach the question of whether *you* have free will until we have answered the much more fundamental question, "What kind of thing are you?" The contrasting criteria cited earlier are founded on differing conceptions of the nature of the self, where the philosophical footing is equally treacherous. Without a shared understanding of what everyone is talking about, it's not surprising that the debate seems to go round and round interminably.

Another barrier to a clear explication of the arguments around whether free will exists is that they are often approached from the direction of their *consequences* for our positions on moral responsibility. If people are not really in control of their actions—if we are nothing more than physical automata, mounting a wonderfully sophisticated but ultimately empty simulacrum of free will—then how can we be worthy of praise or blame? How can we defend judgment or punishment? The stakes here could not be higher. The idea of moral responsibility is the foundation not only of our legal systems but also of all our social interactions. We are constantly thinking about what we should or shouldn't do in any given circumstance and probably spend even more time thinking about what other people should or shouldn't do (or should or shouldn't have done).

But tying the discussion of free will to the issue of moral responsibility muddies the waters. Questions of moral responsibility are crucially important, of course, but they are confounded by all kinds of additional issues: the nature and origins of our moral sensibilities, the evolution of moral norms, the legal philosophies underpinning our justice systems, and the complex and innumerable pragmatic decisions that societies and individuals have to make to keep our collective existence stable. Asking what kind of free will *we want* that will let us maintain our positions on moral responsibility can become almost a theological exercise in motivated reasoning. It means we are looking for a palatable answer instead of trying to understand what really *is*. It is coming at the question from the wrong end, picking an answer we like and seeing what edifice of arguments we need to build to support it. Instead, I would like to know what kind of free will *we actually have*.

## Back to the Start

It's fashionable these days to claim that "free will is an illusion!": either it does not exist at all, or it is really not what we think it is. I am not willing to give up on it so easily. In this book I argue that we really *are* agents. We make decisions, we choose, we act—we are causal forces in the universe. These are the fundamental truths of our existence and absolutely the most basic phenomenology of our lives. If science seems to be suggesting otherwise, the correct response is not to throw our hands up and say, "Well, I guess everything we thought about our own existence is a laughable delusion." It is to accept instead that there is a deep mystery to be solved and to realize that we may need to question the philosophical bedrock of our scientific approach if we are to reconcile the clear existence of choice with the apparent determinism of the physical universe.

But if we want to solve this mystery, humans are the absolute worst place to start. It is a truism in biology to say that nothing makes sense except in the light of evolution—and this is surely true of agency. Instead of trying to understand it in its most complex form, I go back to its beginnings and ask how it emerged, what the earliest building blocks were, and what the basic concepts should be. How can we think about things like purpose and value and meaning without sinking into mysticism or vague metaphor? I argue that we can do so by locating these concepts in simpler creatures and then following how they were elaborated over the course of evolution, increasing in complexity and sophistication as certain branches of life developed ever-greater autonomy and self-directedness.

Indeed, before tackling the question of free will in humans, we have a much more fundamental problem to solve. How can any organism be said to *do anything*? Most things in the universe don't make choices.

Most things—like rocks or atoms or planets—don't do anything at all, in fact. Things happen to them, or near them, or in them, but they are not capable of action. But *you* are. You are the type of thing that can take action, that can make decisions, that can be a causal force in the world: you are an agent. And humans are not unique in this capacity. All living things have some degree of agency. That is their defining characteristic, what sets them apart from the mostly lifeless, passive universe. Living beings are autonomous entities, imbued with purpose and able to act on their own terms, not yoked to every cause in their environment but causes in their own right.

To understand how this could be, we have to go right back to the beginning, to the very origins of life itself (see Figure 1.3). This is the trajectory that I sketch out in this book.

From the chemistry of rocks and hydrothermal vents—the chemistry of the evolving planet itself—life emerged as systems of interacting molecules, interlocked in dynamic patterns that became self-sustaining. The ones that most robustly maintained their own dynamic organization persisted, replicated, evolved. They became enclosed in a membrane—a tiny subworld unto themselves—exchanging matter and energy with their environment while protecting an internal economy and reconfiguring their own metabolism to adapt to changing conditions. They became autonomous entities, causally sheltered from the thermodynamic storm outside and selected to persist.

A new trick was invented: action, the ability to move or affect things out in the environment. Information became a valuable commodity, and mechanisms evolved to gather it from the environment. With that came the crude beginnings of value and meaning. Movement toward or away from various things out in the world became good or bad for the persistence of the organism. These responses were selected for and became wired into the biochemical circuitry of simple creatures.

As multicellular creatures evolved, a class of cells—neurons—emerged that specialized in transmitting and processing information. Initial circuits acted as internal control systems, designed to coordinate the various muscles or other moving parts of the multicellular animal, defining a repertoire of useful actions. At the same time, neurons coupled

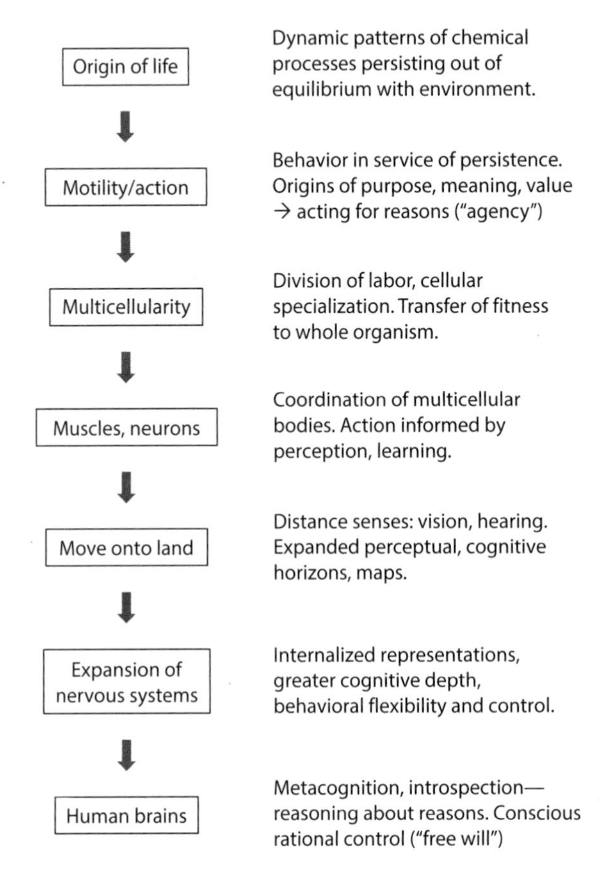


FIGURE 1.3. The evolution of agency and free will. The major stages of evolution of perception, cognition, and behavioral control.

various sensory signals to specific actions in this repertoire, hardwiring adaptive instincts for approach or avoidance.

With the elaboration of the nervous system, this kind of pragmatic meaning eventually led to semantic representations. Perception and action were decoupled by layers of intervening cells. Instead of being acted on singly and immediately like a reflex, multiple sensory signals could be simultaneously conveyed to central processing regions and operated on in a common space. Circuits were built that integrated, amplified, compared, filtered, and otherwise processed those signals to extract information about what was out in the world and what that meant for the organism. More and more abstract concepts were extracted—not just about things but also types of things and types of relations between them. Creatures capable of understanding emerged.

Meaning became the driving force behind the choice of action by the organism. That choice is real: the fundamental indeterminacy in the universe means the future is not written. The low-level forces of physics by themselves do not determine the next state of a complex system. In most instances, even the details of the patterns of neural activity do not actually matter and are filtered out in transmission. What matters is *what they mean*—how they are interpreted by the criteria established in the physical configuration of the system. Animals were now doing things for reasons.

That causal power does not come for free: it is packed into the organism through evolution, through development, and through learning. It is encoded in the genome by the actions of natural selection. And it is embodied in the physical structure of the nervous system in the strength of neuronal connections that express functional criteria in relation to a hierarchy of aims of the organism. There is nothing here that violates the laws of physics; it just demands a wider concept of causation over longer timeframes and an understanding that the dynamic organization of a system, which encodes meaning, can constrain and direct the dynamics of its component parts.

And yes, your actions are at any given moment constrained by all those prior causes. Yet you could just as well say, more positively, that they are *informed by* prior experience. That is precisely the property that sets life apart from other types of matter: living things literally *incorporate* their history into their own physical structure to inform future action. For those who would argue this impinges on the freedom of the self to decide at any moment, I counter that it is this very process that enables the self to exist at all. There is no self in a given moment: the self is defined by persistence over time.

And though you are configured in a certain way that reflects all this history, you are not hardwired. We humans have the remarkable capacity for introspection and metacognition. We can inspect our own programming, treating goals and beliefs and desires as cognitive objects that can be recognized and manipulated. We can think about our own thoughts, reason about our own reasons, and communicate with each other through a shared language. We can access the machine code running in our brains by translating high-level abstract concepts into causally efficacious patterns of neural activity. This gives a physical basis for how decisions are made in real time, not just as the outcome of complex physical interactions but also *for consciously accessible reasons*, and it provides a firm footing for the otherwise troublesome concept of mental causation.

So, if you want to know what kind of thing you are, you are the kind of thing that can decide. Not just a collection of atoms pushed around by the laws of physics. Not a complex automaton whose movements are determined by the patterns of electrical activity zipping through its circuits. And not an NPC, unknowingly driven by its programming. You are a new type of thing in the universe—a self, a causal agent. In the game of your life, you are Player One.

What follows is thus a full-throated defense of the idea of free will. Despite many claims to the contrary, the latest science—whether physics, genetics, neuroscience, or psychology—does not in fact imply that we have no choice or control over our actions. It's true that we are learning more and more about the mechanisms underlying our cognition and behavior—from neural systems and circuits down to the level of cells and molecules or even atomic physics. But even though our cognitive systems have a physical instantiation, their workings cannot be *reduced to* this level. We are not a collection of *mere mechanisms.* As we will see, the nervous system runs on meaning.

The fact that our capacities for cognitive control are grounded in definable biological systems does, however, have important implications for issues of moral and legal responsibility, though these are notably more subtle than the typical absolutist framing. I return to consider these and related issues in the final chapter.

Along the way, I offer a perspective on life that centers agency as its defining characteristic. What distinguishes living organisms is that *they do things, for reasons.* They behave in a truly purposeful manner. This is not an illusion or just a convenient way of talking or thinking about them: it's the right way of thinking about them. Causation does not all bubble up from the bottom, nor is it all instantaneous. The way things are organized can and does govern the way complex systems behave. Living organisms accumulate causal power by coming to embody aspects of their history in their own structure, either through evolution or over the course of their individual lifetimes. The story of agency is thus the story of life itself, and that is where we begin, in chapter two.

# 2

# Life Goes On

For a long time, nothing in the universe did anything. There was a lot going on, to be sure: the early universe was a roiling, turbulent maelstrom of matter and energy—interacting, exploding, colliding, transforming—as it expanded. It's not that nothing happened—it's that nothing in the universe could be said to be *doing* any of it. And then, at some point, actors emerged on this stage. Simple lifeless components were somehow assembled into forms that held themselves apart from these general happenings and instead acted on the world. Entities that do things came into existence.

How did this transition occur? What really is the distinction between living and nonliving? Let's consider a simpler question first: What is the distinction between living and dead? To take an illuminating example from Monty Python, what is the difference between a parrot and an ex-parrot? In a famous sketch, a customer complains to the owner of a pet store that the parrot he bought is dead. The owner protests that it's "just sleeping," to which the customer (a wonderfully exasperated John Cleese) eventually responds: "This parrot is no more! He has ceased to be! 'E's expired and gone to meet 'is maker! 'E's a stiff! Bereft of life, 'e rests in peace! If you hadn't nailed 'im to the perch 'e'd be pushing up the daisies! 'Is metabolic processes are now 'istory! 'E's off the twig! 'E's kicked the bucket, 'e's shuffled off 'is mortal coil, run down the curtain, and joined the bleedin' choir invisible!! THIS IS AN EX-PARROT!!"[1]

1. Monty Python, *Dead parrot*, season 1, episode 8, http://montypython.50webs.com/scripts/Series_1/53.htm.

The ex-parrot has "ceased to be," implying that being alive was not just a state or a property of the former parrot but an activity that it *was doing*. When it died, the stuff it was made of did not change: it just stopped doing what it had been doing. Its "metabolic processes were now history." Staying alive is hard work. It takes a lot of energy to keep replacing all the bits one is made of and to keep them all organized, just so. And when those processes stop, that structure rapidly crumbles—the bits that had been corralled into a defined organization are set free to wander off on their own. Ashes to ashes, dust to dust. It's back to pushing up the daisies for our ex-parrot.

This central idea—that living things have to work hard to keep themselves organized—was nicely articulated by the physicist Erwin Schrödinger in a series of lectures he delivered in Trinity College Dublin in 1943 called "What Is Life?" The short book he published of the same name became a landmark in biology—for a time at least—inspiring people like Watson and Crick in their thinking about the structure of DNA.

Schrödinger took a physicist's view of the question, recognizing that living things, to stay organized, must keep working against what is known as the second law of thermodynamics. This law can be stated and interpreted in lots of different ways, but the gist of it is that systems left to their own devices will tend to become disordered. Though he was not talking about it directly, W. B. Yeats captured the sense of it:

> Things fall apart; the centre cannot hold;
> Mere anarchy is loosed upon the world.[2]

Unless there is some other force keeping them in order, all atoms tend to get mixed up and randomly dispersed. That is because—when they are not constrained in a solid at least—atoms tend to jiggle around and bump into each other and move. With all that jostling going on and no one in charge to maintain order, any structure that was present rapidly disappears, and the atoms tend to spread out evenly into a homogeneous mush: what in physics is known as thermodynamic

2. William Butler Yeats, *The second coming* (Dublin: Dial, 1920).

equilibrium. That's why, when you pour cream into your coffee, you would be surprised if all the cream stayed on one side of the cup, just as you'd be surprised if you put an ice cube into a hot drink and it didn't melt. Things tend to even out.

One way to think about the second law is as a statement about relative probabilities. If someone threw a deck of cards in the air and you picked the cards up at random, you would be very surprised if they ended up all in order, from the deuce to the ace of clubs, and then the same in diamonds, hearts, and spades. In fact, any *specific* random order is just as unlikely as that one. The difference is there is a practically infinite number of ways for the deck not to be in order and only one way for it to be in order. So, just numerically, the ordered outcome is vanishingly unlikely to arise by chance.

The same is true for the organization of atoms in your body. There is an infinite number of other ways that those atoms *could be* arranged, but they are not in any of those ways: they are arranged in a way that makes you. From that point of view, you are a super unlikely arrangement of matter.

But there's more to it than that. When our doomed parrot became an ex-parrot, there was no wholesale alteration of the arrangement of its constituent atoms. That would happen eventually, as its corpse decayed and its physical constituents dispersed, but the immediate transition from life to death involved something else—a cessation of the inner dynamics. Blood stopped flowing, nerves stopped firing, the metabolic pathways inside each cell stopped cycling.

That lets us recognize something vitally important: life is not a state; it is a process. You are not just alive, you *are living*—that is an activity you are doing, that each of your cells is doing. You are more than the pattern of physical matter that makes up your body right now: you are that pattern persisting through time. The individual atoms and molecules and cells that make up the pattern are being turned over and exchanged with the outside world all the time, but the pattern remains.

Life is thus like a storm or a tornado or a flame: none of those things is *made of* the physical atoms or molecules contained in it at any moment. Those elements will be replaced in the next moment. The storm is the ongoing process that is organizing or constraining all those molecules into a higher-order pattern. It's the physical relations between all the elements that is maintained. The difference is that storms or flames fairly quickly blow or burn themselves out, but life does not: life goes on.

## Life in Chemistry

Life could in theory be made of all kinds of things, but on earth it is constituted in chemistry; specifically, the chemistry of carbon. Carbon atoms can form stable bonds with other elements such as hydrogen, nitrogen, and oxygen, but especially with other carbon atoms. Because each carbon atom has four sites where such bonds can form, it can form rings or long chains that other elements can be attached to, thus acting as a scaffold for really big, complex molecules.

Our cells are made out of those kinds of "macromolecules"—proteins and lipids (fatty molecules) and carbohydrates and nucleic acids like DNA and its cousin RNA. Even in a single cell it takes quite some effort to make all those things and to organize them just so. That is achieved through what we call metabolism: interlocking cycles of chemical reactions, catalyzed by networks of enzymes, all finely tuned to keep the cell in good working order.

This constant hum of chemical activity needs a supply both of raw materials and fuel. Just as we cannot have a perpetual motion machine, the chemical reactions in a cell cannot keep running without some intake of energy. That is another way to frame the second law of thermodynamics: no process can be 100 percent efficient. Some energy is always lost (or, more precisely, is transformed from useful energy into useless energy). So, living organisms have to take in some source of "free energy" that can be used to do work, and they give out to the environment a much more disordered form of energy as heat. In keeping themselves organized, they thus contribute to the relentless increase in disorder of the universe as a whole.

Living things thus persist through time, but in the exact opposite way that things like rocks persist. Rocks can exist for eons because they are chemically inert and unreactive, as well as resistant to physical forces.

The atoms in a rock may have been there for millions of years. Their stability comes from stasis. The stability of life, in contrast, comes from constant flux. The individual atoms and molecules are constantly turning over: what is maintained is the network of chemical reactions in which those molecules are involved.

A living cell is self-sustaining as a whole integrated system. The enzymes catalyzing the network of chemical reactions do not only generate new stuff. They are tuned in such a way—that is, their chemical affinities and kinetics and relative levels are set—such that they collectively maintain the relations between all the elements in this dynamic network. They have to be tuned this way, for a very simple reason: if they were not, the "system" wouldn't exist long enough for us to be talking about it.

## From Geochemistry to Biochemistry

How could a living organism—a cell with this kind of complicated, self-sustaining metabolism—arise? Darwin speculated that life might have begun in "a warm little pond" somewhere, which brings to mind a quaint and peaceful garden—an idyllic scene where the earliest life-forms were gently incubated. It seems more likely, however, that life was forged in what might seem at first like a much more forbidding environment—at the chemically reactive interface between the shifting rocks of the earth's crust and the salty oceans of the young planet.

The problem with the idea of complex molecules arising from some kind of a primordial soup is that there is nothing to drive the process. It's very unlikely to happen spontaneously: it needs a source of energy. But for energy to be useful it has to be ordered or unevenly distributed, like the built-up charge on two sides of a battery. If there is no difference or gradient in charge between the two sides, the battery is flat. In the warm little pond, there would be no such energy gradient—everything would be evenly spread out.

You might think that sunlight could offer such a source of ordered energy, but the ability to safely capture this energy arose late in the evolution of life and requires all kinds of complicated molecular machinery.

On the contrary, it seems likely that life arose in the darkest depths of the ocean, making use of an energy source from the earth itself and one that still powers all known life today.

It turns out that all organisms—from bacteria and amoebas to plants and animals—use a gradient of hydrogen ions as a mechanism to power the production of cellular energy stores. Hydrogen is the simplest element: its nucleus comprises a single proton and a single neutron, and a single lonely electron orbits it. The positive charge of the proton is normally balanced by the negative charge of the electron. But when a hydrogen atom reacts chemically with other elements, it can "donate" the electron to them, leaving a positively charged hydrogen ion, symbolized as H+.

Just like in a battery, if you have lots of these ions on one side of some kind of barrier and few on the other side, then you generate a motive force—the hydrogen ions will, if given the chance, flow down this gradient until their numbers are equal on each side and the driving force is eliminated. The energy in that flow can be captured to power something else, in the same way a hydroelectric dam captures the potential energy of water (when it's high on one side and low on the other) to drive turbines and create electricity.

Cells use a similar mechanism to create an energy-carrying molecule called ATP (adenosine triphosphate). As the name suggests, this adenosine molecule, which happens to be one of the four basic components of RNA and DNA, has three phosphate groups (a phosphorus atom bound to four oxygen atoms) attached to it. Because it takes energy to stick those phosphates on, when the bonds are broken they can release energy that can be used to do other things like powering chemical reactions throughout the cell.

Simple life-forms like bacteria have a protein that sits in their cell membrane, called ATP synthase (because it synthesizes ATP). It acts as a channel through which H+ ions from the outside can pass to the inside. As they pass through, they power the mechanism of the ATP synthase, which takes a molecule of adenosine with only two phosphate groups attached to it (ADP) and adds a third (to make ATP). The reason that H+ ions flow into the cell is that their concentration outside is greater than the concentration inside. This doesn't happen by accident: that gradient has to be actively maintained by other proteins that pump the H+ ions back out again.

But even before there were cells with membranes and with insides and outsides, this kind of gradient of H+ ions existed and could have been used similarly to drive chemical reactions to generate complex organic molecules. Hydrothermal vents at the ocean floor may have provided exactly the conditions needed for life to arise. These vents are spots where seawater percolates down through porous rock to depths far below the earth's crust, where it reacts with rocks newly exposed by the movement of tectonic plates. This heats up the water and changes its chemical composition, suffusing it with mineral elements.

This warmed water then returns to the surface, back through this network of porous rock, which is now rich in dissolved hydrogen and carbon dioxide and is highly alkaline. Because the sea itself is slightly acidic, this sets up a gradient of H+ ions between the water rising up in the micro-channels of the rock and the water in the sea. These ions could be conducted by minerals like iron sulfide in the rocks, allowing them to flow across this natural barrier. The free energy from this flow could then encourage the reaction of hydrogen with carbon dioxide to form larger organic molecules (see Figure 2.1).

Crucially, these molecules could have become concentrated within the natural micro-compartments of the rock, rather than diffusing into the open ocean. The proximity of all these molecules and the continuing availability of free energy meant that additional chemical reactions were much more likely to occur, generating ever more complex macromolecules, including nucleic acids, amino acids and peptides (the building blocks of proteins), and complex carbohydrates and lipids. Organic molecules could thus have arisen from the chemistry and geology of the evolving planet itself.

This brings us to a stage where lots of interesting organic chemistry was happening in the rocky nooks and crannies around hydrothermal vents: sets of chemical reactions were feeding off each other, maintained in a dynamic state that was far from equilibrium by a constant flow of free energy and raw materials. But we're not quite to the point where we could identify any thing as actually being alive: there is nothing you would call *an entity*. That took a new invention—a homegrown barrier

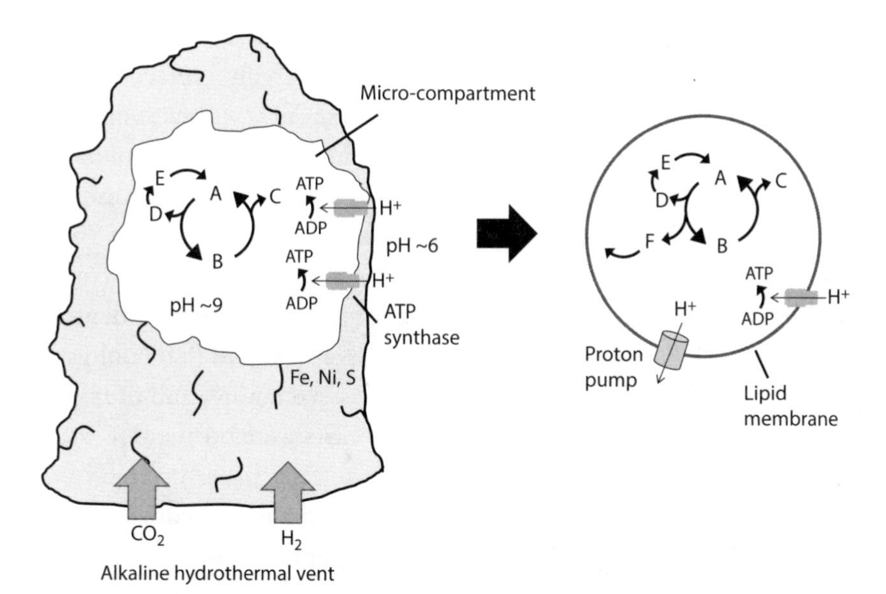


FIGURE 2.1. From geochemistry to biochemistry. Early life may have arisen in alkaline hydrothermal vents, where the raw materials for organic chemistry were concentrated in rocky micro-compartments, with access to free energy in the form of a proton ($H^+$) gradient, inorganic catalysts (iron, nickel, and sulfur) to catalyze biochemical reactions, and a warm temperature conducive to the formation of macromolecules. All these factors could have enabled self-sustaining sets or patterns of chemical reactions to arise and be stabilized. Over time, those chemical reactions may have included production of a lipid membrane. The evolution of proteins that could pump protons out of the membrane would have allowed these proto-cells to generate their own proton gradient to power their internal biochemistry, leading to the emergence of free-living life-forms that could actively maintain their internal dynamics out of thermodynamic equilibrium with the environment.

to separate all those chemical reactions from the environment and allow an autonomous organism to insulate itself from the rest of the world.

## A Storm in a Bubble

That barrier was probably formed from complex lipids: fatty molecules that are chemically hydrophobic (they do not mix with water). Certain types of "fatty acids" have a bit that's hydrophobic and a bit that isn't. At high enough concentrations, these tend to come together to form a double layer, with the hydrophobic bits all huddling together on the inside and the other ends of these molecules happily interacting with any surrounding water. As these aggregates grow in size, they spontaneously form into little vesicles or bubbles, keeping the watery environment on the outside but also enclosing a little watery world on the inside.

Inside that membrane, an entire chemical economy could then emerge. Concentrated in that tiny space, the network of chemical reactions could drive the formation of more and more complex macromolecules, including large proteins or RNA molecules that could act as enzymes to speed up chemical reactions even more, and the lipid molecules that form the membrane itself. Cells were born not as bags of stuff but as bags of activity.

That cellular economy could not be totally closed off from the outside world, however—it still would need a supply of raw materials and energy and some way to get rid of waste products. Otherwise, all those chemical reactions would just go to equilibrium. That import–export business is carried out by protein molecules that span the cell membrane and act as channels or transporters for different kinds of chemicals.

Cells still bear the memory of their rocky origins by using a gradient of H+ ions to generate ATP, which is then available as a source of energy throughout the cell. H+ ions flow into the cell through the ATP synthase protein, which spans the cell membrane. But instead of needing to rely on the unique geology of hydrothermal vents to generate this gradient, cells are able to generate it themselves by actively pumping H+ ions back out. The energy required for this activity comes in turn from stripping electrons from other chemicals taken into the cell; that is, food.

With these mechanisms in place, cells don't need to stay anchored to a naturally occurring H+ ion gradient to power all this organic chemistry. They just need to take in some food—other organic molecules that have some stored energy in their chemical bonds that the cell can release to power the H+ ion pump and make its own gradient of H+ ions. At the same time, the food supplies the raw materials that the cell needs to make its own complex macromolecules. Crucially, the food does not have to be distributed in any special way: it just has to be available. Life was thus set free, as geochemistry had become biochemistry.

The invention of the cell membrane marked another key milestone in the evolution of life: it meant that what was going on inside a cell was physically and chemically *and causally* insulated from what was happening outside. This made possible a break, a discontinuity in the normal flow of cause and effect. All kinds of things could be happening just outside the cell that would have no effect at all on what was happening inside. Only certain types of chemicals could get through that barrier, mainly the ones that the cell needed to put to its own uses. Now the cell was a thing that stood apart from the world, keeping itself going, maintaining all those metabolic reactions, persisting through time as a distinct thing—it was *an entity*.

## Self-Reference

This kind of proto-life would have a degree of independence—of freedom—from the environment. But that freedom is fragile. The metabolism of a cell could be maintained under normal conditions through balancing the relative levels and kinetics and affinities of all the enzymes and molecules that make it up. But all those parameters have to be tuned to be compatible with each other and to form one big viable network for the whole system to keep going. This makes the cell vulnerable to disturbances or changes in conditions.

A good way to get over that vulnerability would be to have some kind of reference or template that could be consulted in the event of some such disturbance, so that the system could reset itself to reinstate all the right parameters and rebalance the network. Ideally, that template would be consulted all the time, not just in emergencies, to keep the system in its normal operating range. It should be part of the cellular economy but not a chemically reactive part; instead, it should be a store of information separate from the dynamic pattern being maintained in the cell and should itself be chemically inert and stable.

That's where DNA comes in—although the current thinking is that DNA's cousin RNA actually came first. DNA and RNA are both "nucleic acids"—macromolecules made of chains of four simple subunits (with DNA and RNA having slightly different chemical compositions of these subunits). In the early stages of the evolution of life, RNA predominated because it provides a necessary stepping-stone between DNA (a stable but inert template) and proteins, which do the business of keeping the cell's metabolism running.

These two elements are so interdependent that it's not obvious how either could have arisen without the other already existing, with all its functions already in place. The DNA codes for the proteins in the cell, but the proteins are required to "read" this DNA code and execute it. RNA may bridge this divide, because it can play the role of both chicken and egg.

Unlike DNA, which just sits there quietly, stably curled in its famous double helix, RNA molecules can be quite active in the cell. Rather than pairing with a matching molecule all along their length, long RNA molecules can fold into complicated three-dimensional shapes and can functionally interact with other macromolecules and even act as enzymes, like proteins. But they can also act as a template, carrying a code to make a protein. Indeed, in our cells, the DNA code has to be transcribed into a molecule known as a "messenger RNA," which carries the information out of the cell's nucleus to where the protein-production machinery is located. This machinery itself involves several other types of RNA molecules with specific biochemical functions.

The "RNA World" hypothesis proposes that in early life-forms RNA was both an information-carrying template and an active player in the biochemistry of the cell. The problem is that RNA is not very stable: it's prone to mutation, as we see in some rapidly evolving viruses that still use RNA as their genetic material. At some point, the information-carrying function was taken over by DNA, in a separation of cellular metabolism from the reference template (see Figure 2.2). DNA is perfectly suited to this role as an information store because it is not only biochemically inactive but also astonishingly chemically stable, which is why it is possible to get DNA samples from the bones of animals or humans who died hundreds of thousands of years ago, or even just from the dirt in caves where they lived.

To sum up, though we do not know the details of how it happened, what I sketched here is our current best guess for how life first evolved. The geochemistry of hydrothermal vents in the young earth provided a

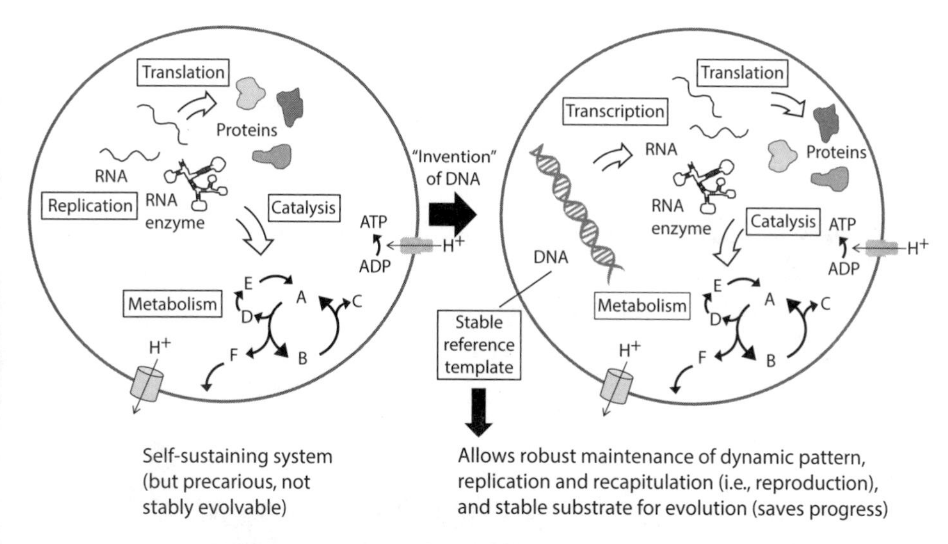


FIGURE 2.2. From RNA to DNA. **(Left)** The earliest cells may have had RNA and proteins capable of catalyzing metabolic reactions. Such systems may be self-sustaining under stable conditions but precarious in the longer run because the entire pattern must be sustained dynamically. **(Right)** The "invention" of DNA provided a stable reference template that is held apart from this dynamic system. This reference encodes the configuration of the cellular components in such a way that it can be used to reinstate a disturbed pattern in an individual cell or recapitulate the pattern in a new cell following replication.

source of both free energy and raw materials, as well as a perfect rocky incubator for organic chemistry to create complex macromolecules. Those molecules became concentrated together and formed interlocking sets of chemical reactions, semi-stable cycles of chemical activity fueled by the energy of the planet itself.

When those reactions became enclosed in lipid membranes, the earliest cells were formed, generating their own energy gradients from food and keeping their internal economy going while being causally insulated from their environment. By this point, these cells could be considered autonomous entities—stable patterns of biochemical processes that have the tendency to keep themselves going, that indeed *strive* to keep going by performing thermodynamic work to stay far from equilibrium and resist the universal drive toward disorder. RNA likely played prominent roles in both early metabolism and information storage but was ultimately supplanted in the latter role by DNA, which provided a much more stable template, underpinning but not chemically involved in the dynamic processes of metabolism.

That's pretty impressive, but if that were all there were to life, then you and I would not be here discussing it right now. Those early life-forms could defy the odds and persist for some time, but to conquer the world, life had to do more: it had to multiply and evolve. That is where the other major advantage of having an information template came into play—it can be copied.

## Go Forth and Multiply

When Schrödinger was writing his short book *What Is Life?* in 1943, it was not known what the genetic material was made of. It was clear that, as cells divided, something physical had to be passed on to provide a template for how each newly born cell should organize itself. The template had to be highly stable—like a crystal—but couldn't be something with a boring, regular structure: it had to have some irregularity to it, so that it could carry some information. Schrödinger speculated therefore that the genetic material must be an "aperiodic crystal."

DNA fits the bill perfectly. Each strand of the double helix is a long chain of subunits (or "bases") of four possible types, designated as A, C, G, and T. These bases have a kind of polarity: they have a "plus" end and a "minus" end. The plus end of one base can be attached to the minus end of another one, and bases can keep on being added in that fashion to make enormously long molecules. It's the irregular, nonrepeating sequence of those bases that carries information, and it does so in two ways. First, there is a code that maps the sequence of DNA bases to the sequence of amino acids in a protein. (Each stretch of DNA that codes for a protein in this way is known as a *gene*; we have about 20,000 such genes in our whole genome.) Second, there is a far less well understood coding that specifies how much of each protein should be made under various conditions.

A single strand of DNA could thus carry the information to configure the biochemistry of the cell. But it is the iconic double helix that provides the means to make a copy. Each strand carries a complementary sequence to its partner. The bonds between the bases along a single strand are not the only biochemical interactions that these bases can make—they can also form weaker bonds with bases on another strand with a remarkable chemical specificity. (You can think of these bonds like the rungs of a ladder, with each strand forming the uprights).

If there is an A on one strand, it will bind with a T on the other, while a C base will bind with a G. This means that if you know the sequence of one of the strands (say, ACGGTTA), you can deduce the sequence of the complementary strand (TGCCAAT). As Watson and Crick remarked in their famous 1953 paper on the structure of the double helix, this structure immediately suggests a means of replication. Just pull the two strands apart, and each one will act as a template on which to assemble the complementary strand by holding individual bases in place, thereby allowing them to be chemically bonded together along the new strand. In this way, two new intact double helixes can be formed.

With this kind of mechanism in place, our newly evolved cells could do something really novel. They no longer had to be content with merely persisting as individual units: they could multiply. The DNA template could be copied, and a single cell could then divide, with a full version of the template delivered to each of the two daughter cells. With the right proteins in place to read the instructions, these two cells could then *recapitulate* the biochemical configuration of all the components. If you are a being capable of this kind of asexual reproduction, you can literally pull in new raw materials from the world outside and make *more you*.

## The Clone Wars

Beings capable of reproduction can do something else that is key to the development of life on the planet: they can *evolve*. A very simple but enormously powerful dynamic drives evolution (in a completely mindless way). In any population of organisms, some individuals are better at persisting and multiplying than others—they are more robust, more stable, more efficient, and faster at making copies of themselves. In a scramble for resources, these different forms effectively compete with each other. The ones making the most of available resources then out-persist and out-multiply the others and increase within the overall population.

That competitive dynamic is at the core of evolution, but it's not enough. The real power of evolution comes from its creativity, its ability to explore new configurations that may be more successful in a given environment and that may even open up new niches where life could take hold and eventually flourish. This creativity arises from the randomness of mutation: changes to the DNA sequence that happen at a very low but steady frequency, usually caused by errors in the copying process.

Some of those changes alter the codescript, resulting in a change in the form and function of some protein or in the regulation of its expression. In the competitive arena, every new mutant is immediately tested to see whether it is better or worse than the previous form and all its neighbors. This kind of blind trial and error thus explores all kinds of possible configurations, at the expense of all the ones that fail or falter. The logic here is entirely circular: if the organism is better at persisting and multiplying, it will persist and multiply more. That's evolution by natural selection in a nutshell.

Simple changes to the DNA sequence were not the only source of change of the genetic material of organisms. Though early organisms did not reproduce sexually, they did have relations of a sort with each other. Simple creatures like bacteria often conjugate transiently through little tubes and transmit a small amount of genetic material to each other. In large populations, genes are often traded in this way like public goods.

Because there is competition to reproduce as quickly as possible, simple organisms typically have very streamlined *genomes* (the term for all the genetic material). That means they tend to get rid of genes that code for proteins that are not immediately useful in their current environment. But as conditions fluctuate, they may need some of those proteins again, and they can get them (sometimes) through this kind of "horizontal" gene transfer between individuals. With all this evolution happening in parallel, different combinations of proteins would arise in different cells, which could open up new possibilities, say, for different kinds of food sources or niches that could be inhabited.

This kind of promiscuous transfer of genes between different cells has advantages, but as the configurations of proteins within any given cell got more complex and selected for how well they worked together, it became less and less productive and more and more risky to simply take up new random genetic material from other organisms. Most such new proteins would be incompatible with and could even disrupt the increasingly complex networks of biochemical processes.

As a result, at some stage in evolution, a transition occurred. The orgy of horizontal gene transfer was curtailed (though not eliminated), and instead most of the transfer of genetic material occurred through cell division—from a mother cell to its two daughters. Cell *lineages* thus became more and more distinct from each other, extending as a continuous pattern through time and still evolving, but with a genetic and historical independence from the other life-forms around them. Though Darwin didn't talk much about these earliest stages of life, this was the true origin of species.

## The Memories of Past Lives

If life is a pattern of activity persisting—doing work to sustain its own organization—that persistence is measured in hours or days or years for individuals but in eons for lineages and species. That extension through time generates a new kind of causation that is not seen in most physical processes—one based on a record of history in which *information* about past events continues to play a causal role in the present.

The stability of DNA makes it an ideal substrate to act as a kind of genetic memory of the experiences of a lineage, with natural selection as the editor. As mutations arise, they are ruthlessly tested in the crucible of competition. Ones that enhance the fit of the organism to its environment will be favored, driving adaptation. Ones that decrease fitness, that undermine this adaptation, will be selected against, so that the individuals carrying such mutations will not persist or reproduce as efficiently (see Figure 2.3).

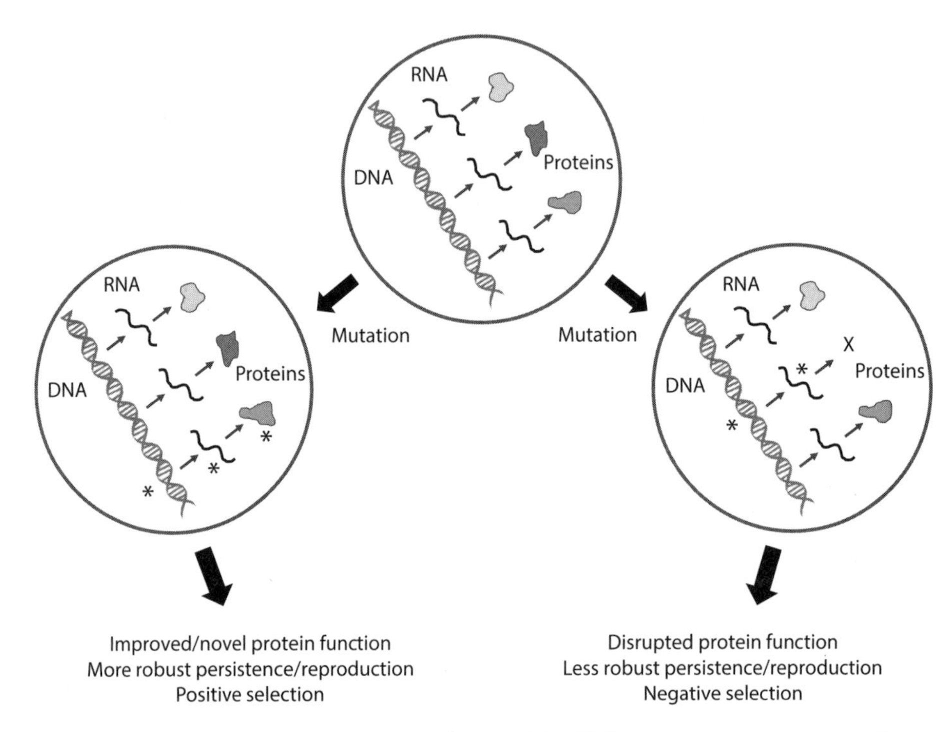


FIGURE 2.3. Positive and negative selection. The DNA sequence acts as a record of the *settings* of cellular dynamics. Mutations can alter that sequence, and natural selection judges the consequences. **(Left)** Mutations that improve the function of a protein or alter it in some way that is adaptive for the cell will be positively selected; that is, individuals that carry that new genetic variant will survive better and breed more than others in the population, so the variant will increase in frequency. **(Right)** In contrast, mutations that disrupt the production or functioning of some protein and thereby impair cellular fitness or adaptiveness will be selected against and removed from the population. The DNA sequence thus comes to reflect the history of the interactions of the organism's lineage with its environment.

This means that the sequence of the genome in a lineage comes to reflect information about the environment and the past experiences of the population of organisms in that lineage, both good and bad. For example, imagine a mutation that alters the function of an enzyme, such that the organism can now metabolize a new source of food (say, a different type of sugar). If that type of sugar is available in the environment, then that mutation may be beneficial and selected for. This sequence of

DNA persists because it carries information about the environment; namely, that this type of sugar can be found in it.

That same process of adaptation happens across the whole genome, reflecting all kinds of aspects of the life histories of individual organisms in the lineage. And the unsuccessful mutations—the ones that *don't* make it through—convey a kind of information too, by their absence. I described the genome as a reference template that carries information about the configuration of all the cell's biochemical components and processes. In a sense, an abstract *model* of the cell is encoded in the genome. And by virtue of the optimizing actions of natural selection, the genome also embodies a model of the environment or, more precisely, of the fittedness of the organism to the environment.

This model allows living organisms to use information about the past to predict and anticipate the future—to adapt to the regularities of their environment by embodying useful functions or responses in their own structure. If we want to understand what happens in a biological system at any given instant, this *historicity* is a key part of the explanation. The sequence of the genome and the resultant structure of the organism *right now* reflect the configurations of its ancestors and whether those configurations promoted persistence or hastened death. In addition to whatever factors are impinging on an organism at any given instant, there are thus causes at work from the distant past, captured in the sequence of the DNA. In a very important sense, there are also causes from the future.

## What's the Point?

The sequence of events described in this chapter might suggest to some a purposive process—as if the point of all that was to generate life. But there's no reason to assume a purpose. Life reflects an unlikely arrangement of matter, but its emergence and the resultant trajectories of its evolution are merely statistical outcomes of the application of the mindless algorithm of natural selection. There is a set of circumstances that led to its emergence but there's no reason for it. If, rather than asking "What is life?," we ask "Why is life?," the answer lies in the question: life is why. It exists because it can. There is no cosmic purpose at play—merely thermodynamic tendencies played out in the particular conditions of our young planet (and of who knows how many others).

But *once life does exist,* everything changes. The universe doesn't have purpose, but life does. Natural selection ensures it. Living organisms are adapted to their environment—retrospectively designed to function in specific ways that further their persistence. Before life emerged, nothing in the universe was *for* anything. But the functionalities and arrangements of components in living organisms are for something: variations that improve persistence are selected for *on that basis,* and ones that decrease persistence are eliminated.

Nothing in this whole arrangement—not the organism, nor any of its components, nor natural selection—is aware of this purpose. Indeed, you might argue that this is not *real* purpose: it's only the system behaving *as if* it had purpose. And its components don't actually have *functions*; they only have physical and chemical properties and tendencies that fit well together to form a whole system that tends to persist over time.

But we can approach the question from the other end and ask, "What would be the difference in the consequences between a system with 'real' purpose and one that only behaves as if it has purpose?" If we put aside the fact that all this apparent design was actually imparted by the blind actions of natural selection, we can see that the result is a *fittedness* of the organism to its niche and a fittedness of each component to the whole system.

Just as importantly, there is a fittedness of the activities of the system to the future outcome of its persistence. Organisms are doing work in order to maintain themselves: all those activities are goal directed (and thus qualify as functions). At least that is the effect, and the pragmatic consequences are no different than if the organism was explicitly designed to have that goal. Things happen because of that goal. The system behaves in ways that advance it. The components of the organism and its internal processes have functions, *with respect to the goal of persistence.* Once you have those properties, what else is needed for something to qualify as a goal or for us to say the whole system has purpose?

And unlike the designed machines and gadgets that surround us in our daily lives, which also have a purpose or at least serve a purpose, living organisms are adapted for the sake of only one thing—their *selves*. This brings something new to the universe: a frame of reference, a subject. The existence of a goal imbues things with properties that previously never existed *relative to that goal*: function, meaning, and value.

In a lifeless universe, things have consequences, but nothing matters. There's nothing with respect to which they could matter. Things just happen, with nobody or no thing trying to make them one way or another. Nothing has meaning or value; nothing is good or bad. But living things do try (to stay alive), and because of that, things matter to them. As we will see in later chapters, meaning and value are the internal currency that drives the mechanisms of decision making and action selection that emerged as life continued to evolve.

From the rocks and sea of our early world, life emerged—organisms that actively maintain their internal states and sustain a degree of causal autonomy from the world around them. The next step in the evolution of agency is the ability of these autonomous organisms to act on the world, to become causes in their own right.

# 3

# Action!

Staying alive is a 24/7 job. Living organisms have to work hard to keep themselves organized. That self-maintenance is not just the mechanical consequence of all the processes inside the cell but it is also the selected *function* of those processes: they work that way precisely because doing so tends to lead to the persistence of the cell. Under steady conditions, those processes just hum along, all the interlocking feedforward and feedback loops ensuring that the whole system stays within an optimal operating range. But when conditions change, the system has to adapt.

And conditions do change for two major reasons. First, the physical systems of the earth are very dynamic, and material is in flux all the time. And second, changing conditions are also caused by the actions of the organism itself and of other organisms in its environment. An organism that eats a certain substance as food may deplete that food source in its surroundings, especially if it is multiplying and creating more hungry mouths that eat the same stuff. At the same time, waste byproducts that are excreted from the cell may build up in the environment, potentially to toxic levels. Thus, organisms have to be prepared for change because they cause change.

Some of that change happens over evolutionary time in an ongoing dance between organisms and the world around them. As organisms multiply, their collective metabolic activities can alter the environment—sometimes on a planetary scale—in ways that create new niches that evolving organisms can then take advantage of. Indeed, we humans only exist because of the wholesale impact of single-celled organisms on the chemistry of the early earth, especially their role in the production of atmospheric oxygen.

Of course, organisms must be capable of changing over much shorter timescales to survive in a treacherous world. This is accomplished by altering the configuration of biochemical pathways, thereby shunting metabolic flux through the system in different directions depending on conditions. For example, some organisms are able to carry out their metabolism in conditions where oxygen is present or where it is absent. These involve divergent metabolic pathways, and you don't want to run both at the same time.

Part of the control of which pathway is used is passive, built into the dynamics of metabolism. In the absence of oxygen, the final chemical reactions in the aerobic respiration pathway cannot be completed, resulting in the buildup of certain metabolites and the shunting of metabolic flux down an alternate route, known as the anaerobic respiration pathway. This pathway is used in many organisms, including some species of yeast—and thank goodness—because the byproducts of this pathway—also known as fermentation—include carbon dioxide and ethanol.

But the biochemistry of these single-celled yeasts is also actively reconfigured under different conditions (see Figure 3.1). Whole sets of genes are, in a coordinated fashion, "turned on or off"; that is, the DNA code for various proteins is either actively read off or not, so that different profiles of proteins are made in each condition. But how does the yeast "know" whether there is oxygen present or not? Dedicated proteins inside the cell are either biochemically activated or deactivated by the presence of oxygen, thus enabling direct monitoring of oxygen levels. In turn, these proteins switch on the respective expression of enzymes needed for aerobic respiration or for fermentation. This is an efficient way to rapidly reconfigure the biochemistry of the cell without having to constantly make enzymes that are not needed.

In this scenario, single proteins—by virtue of altering their activity depending on whether they have bound some chemical or not (in this case, oxygen)—are effectively carrying out a kind of logical operation: if A, do X; if B, do Y. They're not thinking about it, of course, but that is the effect, and it's built right into the design of the molecule. This type

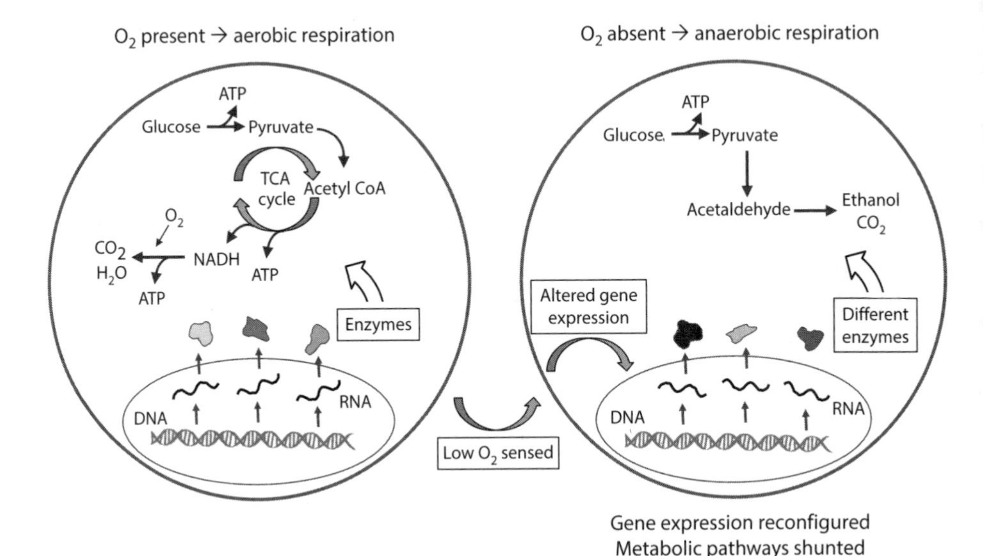


FIGURE 3.1. Reconfiguring metabolism. One way that an organism can improve its chances of persisting is by reacting internally to changing environmental conditions. Many single-celled organisms can switch their metabolism between an aerobic regime **(left)** when oxygen is present to an anaerobic regime **(right)** when it is absent. This switch is mediated by receptors that sense oxygen levels and regulate different patterns of gene expression to drive the change in metabolic dynamics.

of mechanism is ubiquitous in living organisms: thousands of such operations are carried out in cells all the time by individual molecules or by groups of them acting in concert. As we will see in later chapters, the same holds true in organisms with complex nervous systems, in which much of the computational work is still carried out at the microscopic level by individual protein molecules.

Another example is found in the gut bacteria *Escherichia coli* (more commonly referred to as *E. coli*). *E. coli* like to eat glucose, a simple sugar that can be easily broken down to power the ATP-producing pathways in the cell. But they can also digest more complicated sugars like lactose, using a dedicated set of enzymes to break it down into glucose (and galactose). When there is no lactose around, there is no point making these enzymes: doing so is expensive and wasteful, and in populations of rapidly dividing cells like *E. coli*, all competing with each other, natural selection doesn't tolerate inefficiency.

So, most of the time these genes are turned off: the DNA code is there, but no messenger RNA or protein is being made from it. Gene expression is clamped by a *repressor* protein that physically binds the DNA sequence of these genes and excludes the enzymes that would otherwise read them into a messenger RNA. That repressor protein also binds something else—lactose. When it's present, lactose binds to this protein and alters its shape. As a result, it can no longer bind to the DNA, and the genes can be expressed. Again, a single protein is carrying out a logical operation (see Figure 3.2).

But the system as a whole has some added complexity. Even if lactose is present, the bacterium does not need to turn these genes on if glucose is also present—it can just happily go on eating the glucose. It only needs those genes when two conditions are met: glucose is absent and lactose is present. So, another protein (called CAP) exists that indirectly senses whether glucose is present and that also binds the DNA and regulates the expression of these genes. CAP is required for these enzymes to be expressed and is inactivated when glucose levels are high.

This whole system—the regulatory regions of the DNA, the two proteins that bind those regions (the repressor protein and CAP), and the molecules that determine whether those proteins are active or not—is thus performing a higher-order computational operation on two conditions: if A and not B, then do X. If lactose is present AND glucose is NOT present, then turn on these genes.

There's nothing really special about this pathway—it just happens to be very well studied and well understood. There are all sorts of similar pathways in any single-celled organism, controlling and coordinating the expression of hundreds or thousands of genes to enable the cell to respond to changes in the chemicals present in its environment and to adapt to varying conditions. In general, cells are busy all the time doing work to adjust their inner economy to respond to a changing environment. That's a good trick, but an even better one is to *anticipate* a change in the environment and reconfigure internal systems to be ready for it before it happens.

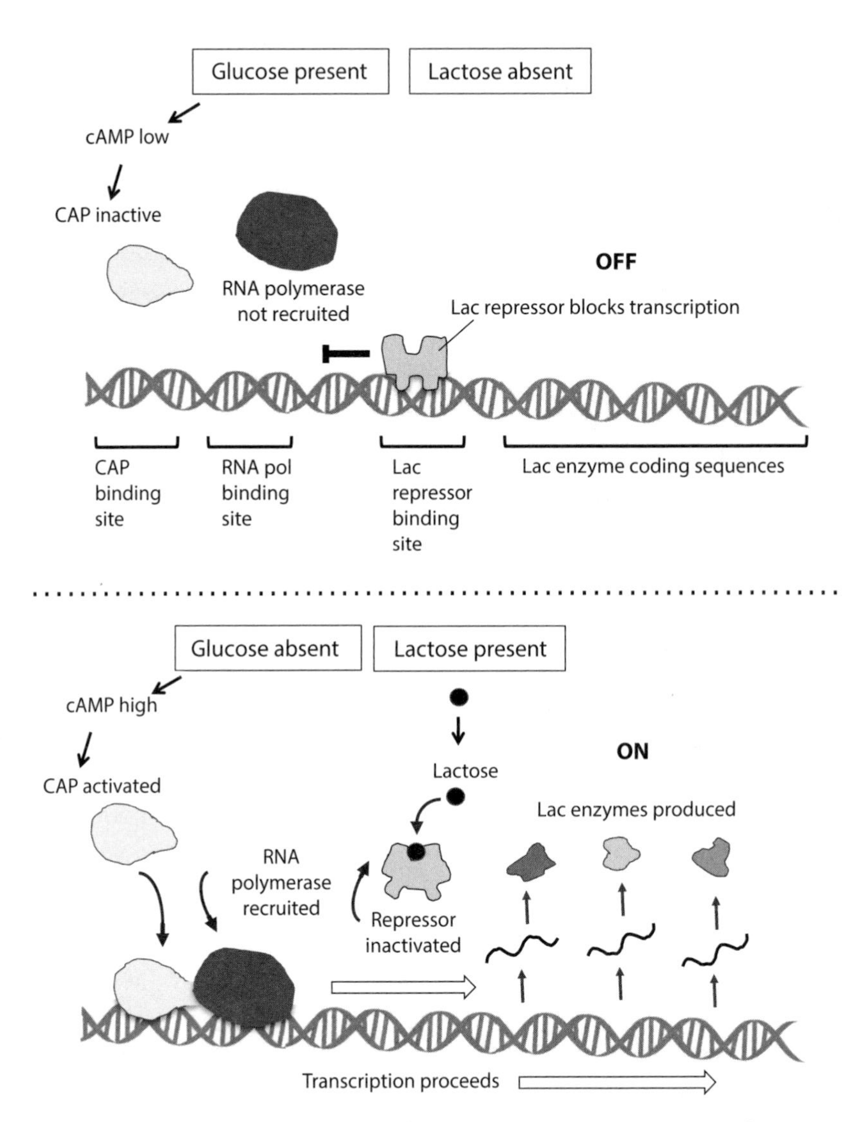


FIGURE 3.2. Regulation of expression of the genes encoding enzymes for lactose metabolism in *E. coli*. This system comprises a logic gate. In conditions where glucose is present OR lactose is absent **(top)**, expression of the lac genes is repressed. Only in the case where glucose is absent AND lactose is present are the genes expressed; that is, the messenger RNA is transcribed and translated into the proteins that allow lactose to be metabolized.

## As the World Turns

Many changes in the environment are, of course, unpredictable, but some are utterly reliable—such as those associated with the regular transitions between day and night. Changes in conditions over the course of the twenty-four-hour rotation of the earth are so important that virtually every organism has evolved a means of anticipating them. This anticipation is achieved through the actions of an internal molecular clock, which runs at approximately a twenty-four-hour period.

Details of the molecular components differ widely among species, but the basic design of the clock is the same: at least two interlocking negative feedback loops regulate the production and slow degradation of specific proteins that, in turn, regulate their own expression or the expression of other components of the clock, usually negatively. As a given protein is produced and its concentration in the cell increases, it shuts off its own production. But then the protein is degraded over a certain timeframe, its concentration drops again, and its expression begins again. Importantly, even though the clock is recalibrated on a daily basis by cues like light or temperature, it can effectively run independently for many days, keeping to that internal twenty-four-hour period.

The important element of this design is that the clock proteins also regulate the expression of lots of other genes in the cell, effectively reconfiguring its internal physiology to prepare it for a change in conditions that is surely coming—as sure as the world keeps spinning. For example, in some single-celled creatures that use sunlight as a source of energy, the genes encoding proteins necessary for photosynthesis are upregulated in the hours *prior to* the rising of the sun; these creatures similarly regulate genes involved in nitrogen metabolism to anticipate the time of day when they will be needed.

By continually selecting individuals that are most adapted to their environment, natural selection effectively packs knowledge about the world into the physical structure of living organisms. This includes matching the dynamics of the organism to the dynamics of the world. Sometimes it's dark and cold, sometimes it's bright and warm, and these periods follow each other with regularity. Sometimes there's oxygen to power cellular respiration; sometimes there's none. Sometimes there's glucose to eat, and sometimes there's lactose. It pays to be ready for these changeable conditions either by responding as they happen or anticipating them.

However, as impressive as all these reconfigurations of cellular physiology are, they are purely internal actions. They let an organism adapt on the fly to changing conditions in its immediate environment, but there are limits to how effective these adaptations can be. If food sources run out or waste products accumulate to toxic levels or the temperature rises too high, no amount of internal reconfiguration is going to save the cell and allow it to sustain itself. In those circumstances, another strategy is needed. If the environment is no longer habitable (or even just no longer optimal), it's time to move.

## Getting Moving

Single-celled organisms have evolved myriad ways of getting around: floating, swimming, crawling, rolling, stretching, sliding, pulling, squeezing, rowing, squirting, and more. The most common motions in aquatic microorganisms involve propulsion through the water by waving tiny hairs on the surface of the cell called cilia or by rotating or undulating one big, long hair-like structure called a flagellum. Other creatures, like amoebas, move by changing their shape: sending out a long extension of the cell membrane called a pseudopod, grabbing hold of some substrate, and then squeezing the rest of the cell into the pseudopod.

We can already see an important difference here: some creatures—especially those with rigid cell walls, like bacteria—can move around but can't move parts of themselves relative to other parts. The whole creature moves, like a boat with oars or an outboard motor, but it maintains its shape. More squidgy creatures can change their shape, reaching out into the world without the whole entity necessarily moving. As we will see later, with the right kinds of control, changing one's shape brings much more diverse opportunities for action in the world that go beyond mere locomotion.

But for now, let's concentrate on moving around. Once organisms had their own means of generating an energy gradient, they weaned themselves from their rocky incubators and could float or swim or crawl away, free to explore and take their chances in the world around them, a world of opportunities but also of dangers. They were no longer at the mercy of a changeable environment: if things got bad, they could just move along. It's easy to see how this capacity would have been selected for as an additional way to help the organism keep its internal dynamics going and increase the likelihood of its persistence.

But the capacity to move immediately raises a key question: Which way should you go? You could move about randomly; in fact, many creatures do just that, floating wherever the ocean currents take them. That is not a bad strategy, especially when lots of such creatures move together. In that scenario, food might become depleted if a great mass of individuals remained in any one area for too long, so keeping moving might be the only way for such creatures to outrun their own appetites. After some point, anywhere else is likely to be better than the spot they have been in for a while.

Random exploration can be a powerful search strategy, especially if an organism has the ability to stop when finding itself in a welcoming spot. This would require it to have some way to sense, for example, the concentration of food in the environment and to put on the brakes when it reaches a place where resources are plentiful; say, by sticking to a rock or some other surface. Conversely, when the cupboard is bare, the organism could pick up stakes and search for greener pastures. We're getting an early hint of the organism doing something, *deciding* something: Should I stay or should I go now?

To be able to answer this question, our little creature can't be completely insulated in its little bubble. It needs to transport food and electrically charged ions across that boundary. But it also needs to know what's going on outside its bubble: it needs information to survive. The organism has to be able to sense things in the environment. In choosing to move or to stay, the organism might, for example, measure the external concentration of a food source and set some threshold for action: when food gets low enough, let's go!

Even better would be if the organism would have the ability to decide *where* to go. In an environment where resources and threats are unevenly distributed, it would pay to move toward the former and away from the latter, rather than just drifting aimlessly around. Such tendencies would certainly increase survival and thereby be selected for. But what would it take to enable these kinds of behaviors? Clearly the organism would need to know what is out in the world, on the other side of its insulating boundary. It has the motors—now it needs the sensors.

One way an organism could sense things about its environment is to physically bring molecules from the outside in—to transport them across the membrane and then react internally depending on their concentration. That might work for food, which it would want to get inside anyway, but is not such a great idea for toxic substances that should be avoided. A way around that problem is to build a sensor: something that sticks out of the cell and detects some molecules or other stimuli, like vibrations or light, and that sends a *signal* into the interior of the cell without actually transporting any substance. This would give the organism an impression of what is out in the world while maintaining the integrity of its boundary with its environment.

## Through a Glass, Darkly

That impression begins with an array of sensors, which are almost always protein molecules that sit in or span the membrane. The part of the protein outside the membrane is typically capable of binding something, let's say a particular chemical molecule. When it does that, the physical conformation of the protein may shift. This can cause a kind of ripple to run through the whole length of the protein, from the part on the outside of the membrane to the part on the inside. A resultant change in the conformation on the inside may then allow binding of some other molecule within the cell, setting off a cascade of internal signals. Alternatively, the protein sensor may act as a channel through the membrane for electrically charged ions like sodium or calcium. When it binds to some molecule, this channel may open, and the flow of ions into the cell can then act as an internal signal.

The important part of this process is that the molecule being sensed is not actually transported into the cell. In most cases, there is no transfer of energy or matter at all. There is just a signal sent—some alteration to the internal biochemistry that carries information, a pattern that is now *about something*. That is, the internal state of the cell is now physically correlated with the presence or absence (or concentration) of some molecule in its environment. This is how smell works in your nose, and it's how simple creatures sense the chemicals around them.

But which chemicals should an organism sense? What kinds of molecules or specific molecules should it be interested in? It should not be concerned with every molecule because there are all kinds of stuff out there that have no relevance for any particular organism. It just needs to be able to sense the stuff that can affect whether it survives or not; for example, whatever kinds of molecules it happens to eat. This food obviously varies for different kinds of organisms, but each needs to be able to detect its own particular sources of food in its environs.

However, the signal that indicates where food is does not need to be the food molecule itself. For example, if you are the type of organism that eats other smaller organisms, digesting them to get at the food molecules inside them, then your goal would be to sniff out where those smaller organisms are. You might, for example, evolve a sensor to detect the waste products that those organisms—your prey—produce. That's how mosquitoes track animals—by sensing the carbon dioxide they expel as a byproduct of respiration.

This is another example of organisms adapting to and, in a sense, embodying the regularities of the world: where there's a high local concentration of carbon dioxide in the air, there's often a big animal that a mosquito can extract a tasty blood meal from. Natural selection builds that expectation into the mosquito's genome and nervous system, and a similar process equips single-celled organisms with a kind of innate knowledge about the world.

Of course, if you happen to be the prey in this equation, it would pay to evolve some means of sensing the predator, so you could take evasive action. That might mean reciprocally sensing a chemical they produce, but you could also detect the vibrations caused by the movement of a larger creature or see a shadow it casts. Many single-celled organisms have sensors for those kinds of stimuli as well: they are able not only to smell but also to hear and see. Anything that gives them a selective advantage—that reliably predicts something useful about the world—may evolve and be retained. Each creature would thus develop an array of sensors adapted to detect resources and threats in its environment.

The sensorium of each organism thus becomes highly particular and selected for just those things of relevance (whether that relevance is direct or indirect) and entirely oblivious to everything else. Natural selection is frugal, and it is expensive to make and operate the molecular machinery of sensation. There's no point wasting energy sensing things that don't matter and that you're not going to do anything about. This is useless information; in fact, it is worse than useless because it would simply generate noise in the system.

We humans can only see a narrow band of wavelengths of light, hear a narrow range of sound frequencies, and smell a limited set of odorants (compared to dogs or elephants, for example). In a sense, we inhabit a selected slice of reality. Single-celled organisms similarly have a highly circumscribed and ruthlessly selective view of what is out in the world. That information is strictly on a need-to-know basis. The organism only needs information that it can—that it *should*—act on.

## What to Do?

The simplest kind of response to a signal about something out in the world is to move toward it or away from it. But how do you know where it is? The answer is not as straightforward as it seems. For an organism to infer the source of a signal and to use that information to decide *where* to move, it needs to measure *relative* levels of the stimulus being detected in different parts of the environment. There are several ways this could be accomplished. An organism could compare signals in space, say from its "front" and the "rear" ends, and favor movement in the direction with the stronger signal. Or, by comparing signals over time, it could determine whether the concentration is increasing or decreasing as it moves in a particular direction and thus whether it should keep going that way or change course.

This is already getting much more sophisticated than a simple threshold response. The signal now is not merely an isolated trigger that the organism responds to automatically. Instead, the organism needs to integrate multiple measurements and compare them effectively to make a model of the outside world. Even at this simplest level of perception it is not just passively responding but is actively involved in making sense of that information to infer what is out in its environment—to build a kind of spatial map of how things are distributed. The organism doesn't really "know" this in any conscious sense, of course, because consciousness has not evolved yet. It may not even separately represent it, but there's still knowledge in the system. The organism's internal biochemical configuration reflects an integrative picture of what is out in the world, and its subsequent behavior is *informed by* that picture.

Different unicellular organisms use different strategies and mechanisms to control their movement in response to information from the outside. In bacteria like *E. coli*, for example, special receptors in the membrane bind to food molecules such as sugars or amino acids and send internal signals to control how the organisms move. *E. coli* move around by rotating a set of long filaments—flagella—that protrude from one end of their rod-shaped body (see Figure 3.3). When they all rotate in the same direction (counterclockwise), they act like one big propeller, moving the bacterium "forward." But when some flagella rotate in the clockwise direction, their concerted action is disrupted, and the bacterium will instead just tumble around.

When the receptors bind to food molecules, they send a signal to the proteins that control the flagella, favoring counterclockwise rotation and ensuring that the bacterium keeps moving in whatever direction it's going. If the receptors are not activated, the opposite signal is sent, causing some flagella to rotate clockwise and the cell to tumble in place. After a few seconds, the cell will switch back to concerted rotation and will head off in another random direction.

This system works because of the involvement of other proteins that rapidly reset the sensitivity of the receptors after their activation. This

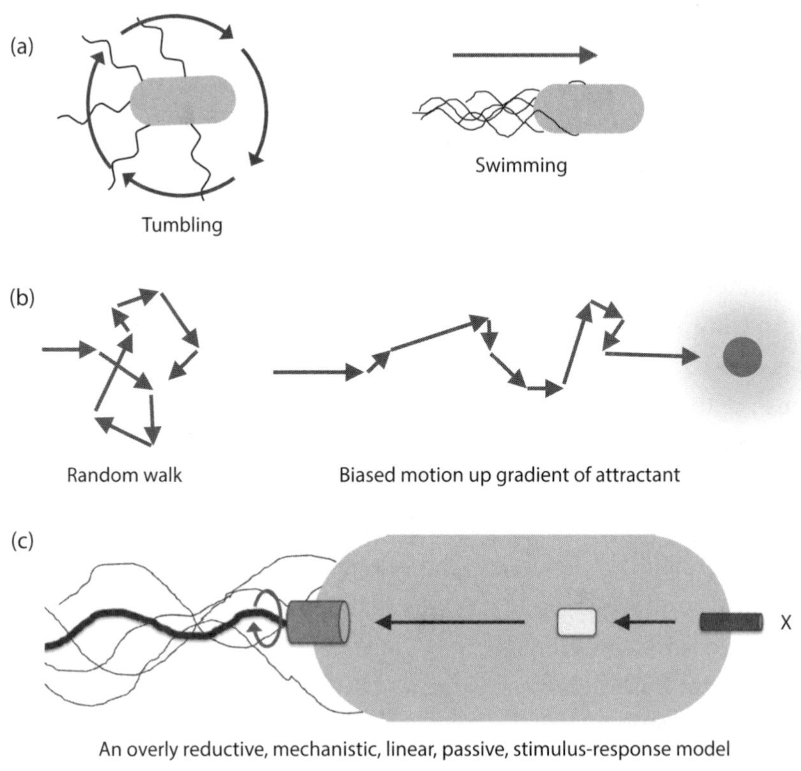


FIGURE 3.3. Bacterial chemotaxis. **(a)** Bacteria like *E. coli* can move in two ways: tumbling in a circle or swimming in a straight line. **(b)** When they detect a food source, they shift from a "random walk" to more directed travel up the concentration gradient, by regulating the direction of rotation of their flagella and the amount of time they spend swimming versus rotating. **(c)** A simple view of chemotaxis sees a simple, linear, stimulus–response mechanism that controls where the bacterium moves **(top)**. A more realistic view sees chemotaxis as a holistic process whereby the active bacterium accommodates to multiple sources of information to derive an adaptive response as an integrated agent **(bottom)**.

effectively allows the cell to compare the concentration of food molecules *over time* as the bacterium moves. If the concentration is increasing, then the activation of receptors occurs faster than deactivation, and the bacterium keeps heading in that direction. If it's decreasing, deactivation wins out, tumbling behavior is triggered, and then the bacterium randomly heads off somewhere else.

The statistical outcome of this behavior is that the bacterium will tend to head toward a source of food (up the concentration gradient), because it will have spent more time swimming in that direction than in any other one. Another crucial aspect of this system is that it modulates *the gain* of the signaling pathway—how strong the signal is relative to the absolute concentration of the food molecules—ensuring that the bacterium remains sensitive to tiny changes in concentration even at very high absolute levels of the food molecule.

Avoiding threats is just as important as approaching resources, of course, and has been well studied in single-celled creatures called Paramecia. These oblong organisms—hundreds of times bigger than *E. coli*—are covered in tiny motile hairs called cilia. These beat in a synchronized fashion to propel the creature forward, like oarsmen on a Viking ship. If the Paramecium bumps into some obstacle or noxious substance as it is moving forward, a signal is triggered that causes the cilia to beat in the opposite direction, prompting an *avoidance reaction*: the beast reverses rapidly, alters its orientation, and then resumes forward motion in another direction. By contrast, if it gets bumped on the tail end—say, by a predator trying to catch it—this prompts the *escape reaction*: the cilia all beat faster and the Paramecium speeds up.

In Paramecia, these internal signals are electrical and are mediated by proteins in the membrane of the cell that are responsive to mechanical disturbance. When the beast gets bumped, these proteins open a pore in the membrane, and calcium or potassium ions from outside the cell rush in, generating an electrical pulse; this process is very similar to that used by nerve cells in more complicated organisms. Different receptors are located at the front and the rear of the organism, controlling which ions flow in and the resulting effect on the cilia. These responses are thus built into the physical structure of the organism.

For some microorganisms, the most salient things in their environment are other members of their own species. This is especially true for the remarkable social amoeba *Dictyostelium discoideum*, also known as slime mold. These organisms start out life as free-living, single-celled amoebas that crawl around in soil, looking for food. Their main prey is our friend *E. coli*, which they detect by sniffing out folic acid that the bacteria excrete. So far, so normal—they are just critters looking for food. When times are good, individual amoebas reproduce by dividing in two, and the population in an area thus increases. It's when the going gets tough that things get weird.

When the bacteria in a region have been depleted, the amoebas release a signal that acts as an attractant for other amoebas, driving them to aggregate. This chemical—called cAMP—is detected by specialized receptors on the cell membrane, which send a signal into the cell that affects the direction of movement. Special proteins inside the cell form a dynamic mechanical structure known as the *cytoskeleton*, which can be rapidly assembled and disassembled in different parts of the cell. Assembly is favored under areas of the cell membrane where cAMP signals are high, causing the cell to first extend and then move in that direction.

Dictyostelium use these signals to find each other. And when they do, they exhibit astonishing behavior (see Figure 3.4). First, as many as a hundred thousand of them aggregate together to form a multicellular superorganism known as a *slug*. This slug itself moves around in response to chemical cues, light, and heat, looking for a suitable place to undergo an even more remarkable transformation. When it has found a good spot, the slug stands on one end and stretches up toward the sky. As it does so, the cells at the bottom form a stalk, and those at the top differentiate into a fruiting body, containing thousands of spores. When these spores are released and dispersed they can germinate into new individual amoebas.

This fascinating slime mold thus presents a wonderful model for understanding not only how cells decide where to move but also the origins of multicellularity, in which individual cells form a collective organism and take on different roles to enable concerted action. We will

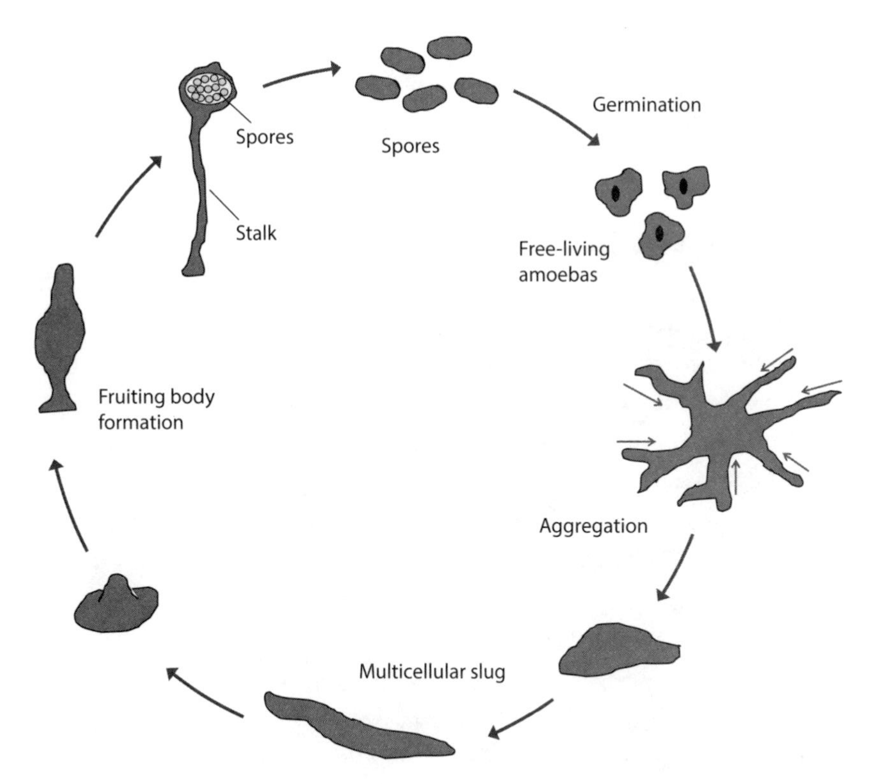


FIGURE 3.4. Dictyostelium life cycle. The slime mold transitions from a unicellular stage, during which individual amoeba seek out resources and divide to increase their numbers, to a multicellular stage where, under more challenging conditions, they coalesce into a motile "slug" and eventually form a fruiting body, releasing spores that will germinate into new unicellular amoebas.

talk more about that process and its opportunities and challenges in the next chapter.

This quick survey of a few species illustrates some of the diverse strategies and mechanisms that single-celled creatures—or multicellular forms—use to navigate through their environment. These organisms are able to control whether they move, in which direction, and in what manner, by coordinating the mechanical components of the cell to select and execute one option from a repertoire of possible actions. They use signals from the environment to inform this process in a goal-directed manner. They are thus making elementary but strategic decisions: Stay or go? Exploit or explore? Approach or avoid?

## Is This Action without an Actor?

Perhaps you think I've gone too far. Is it really correct to say that these simple organisms are selecting an action? Is the organism—as an entity—really *doing something*? Or is it just that it contains some mechanisms that, when triggered by some external stimulus, result in the movement of the whole structure? Is the organism just a complicated machine, driven by physical forces impinging on it and playing out within it, or is a different kind of causation at play?

First, let's ask ourselves what would justify our characterizing any of these single-celled organisms as an agent—a locus of causation—in the scenarios outlined earlier. A primary condition might be that its behavior is not completely determined by any given physical stimulus. If every time it encountered some object, it always behaved in exactly the same way, we would be correct in perceiving a simple stimulus–response machine at work. There would be little justification for granting any causal responsibility to the organism as a whole in determining the outcome.

As it happens, there is substantial variability in the behavioral outcomes arising from any given stimulus. This is primarily caused by the nature of the machinery, which is made from wetware, not hardware. All those tiny components are jittering about in the cell all the time, binding other molecules and letting them go, with the precise number of protein molecules fluctuating as some degrade and new ones are produced. The cell is never the same from moment to moment: it is a complex dynamic system, and its response to even the exact same stimulus may be affected by the precise state it happens to be in at the time.

This variability highlights a crucial fact: the system is not sitting idle, passively waiting for some trigger to provoke a specific reaction. It is a hive of ongoing internal activity. Signals from the outside are accommodated into this web of dynamic processes in a way that biases activity, rather than directing it. Indeed, many of the responses we have seen are implemented statistically—by altering the relative likelihood of a bacterium swimming straight or tumbling, for example.

Moreover, although we discussed behaviors in response to specific signals in isolation, in reality this is not how the world works. And it is not how organisms perceive it or respond to the environment. Even single-celled creatures are capable of integrating multiple signals at once and altering their behavior accordingly.

We already saw an example of that in *E. coli*'s ability to reconfigure its biochemistry to be able to digest lactose, but only when two conditions are met: glucose is absent and lactose is present. These conditions are mediated by two independent signaling mechanisms integrated at the level of regulation of the expression of the genes encoding the lactose-digesting enzymes. We saw how the relative concentration of a food source is computed over time to guide the direction of movement; the bacterium effectively makes an inference about the distribution of the signal in the environment. But bacteria also respond to many other factors in their environment representing opportunities or threats, and all this information must be integrated to guide behavior in the way that is most adaptive to the situation as a whole.

Paramecia similarly respond to and simultaneously integrate many kinds of stimuli (food, heat, light, etc.) to optimally direct their movement. They are, for example, attracted to molecules released by bacteria (like folic acid, glutamate, or ammonium) but repelled by other chemicals (like ATP, for example, which is released at high concentrations from dying cells, thereby signaling a generally inhospitable environment). Paramecia prefer slightly acidic to alkaline conditions and are attracted to carbon dioxide, which is released by other Paramecia, thus driving a tendency toward social clustering. These creatures have a rich and complex ecology, including a social life, which drives a need for similarly complex behavior.

Moreover, both *E. coli* and Paramecia adjust their signaling responses to which specific food molecules are close by, the temperature, the ionic composition or pH of the environment (how acidic or alkaline it is), and the density of other cells. These adjustments keep the system in an optimal operating range. *E. coli* achieve this state by altering the levels or relative composition of proteins in their sensory arrays, which tunes their sensitivity to various chemicals. Paramecia make similar adaptations, altering the speed at which channels in the membrane conduct electrical ions or the sensitivity of the mechanisms that regulate the cilia to these internal electrical signals. These adaptive mechanisms also enable simple kinds of learning such as habituation or sensitization, which make it possible for the organism's response to a signal to change based on its recent experience.

Although their behaviors appear simple from the outside, these single-celled creatures are thus far from being passive stimulus–response machines. Their response to a given signal depends on what other signals are around and on the cell's internal state at the time. These organisms infer what is out in the world, where it is, and how it is changing. They process this information in the context of their own internal state and recent experience, and they actively make holistic decisions to adapt their internal dynamics and select appropriate actions.

This represents a wholly different type of causation from anything seen before in the universe. The behavior of the organism is not purely driven or determined by the playing out of physical forces acting on it or in it. Clearly, a physical mechanism underpins the behavior, which explains how the system works. But thinking of what it is doing—and why it is doing it—in terms of the resolution of instantaneous physical forces is simply the wrong framing. The *causation* is not physical in that sense—it is informational.

## Information and Meaning

Before we go on, let me discuss what I mean by information, why it is useful for organisms to have it, and how it can have causal power in a physical system. First of all, information is physical; that is, it has to be carried in the physical arrangement of some kind of matter. It doesn't just float around in the ether but has to be instantiated in some physical medium. That is literally what the phrase *to in-form* means. As a result, the idea that information can have causal power in a physical system should not be so outlandish: it is the same as saying that the way a system is organized constrains how it evolves over time.

The word "information," however, has multiple meanings, both colloquially and technically, which can create confusion. The most commonly used technical sense of the word was developed by Claude Shannon working at Bell Labs in New Jersey and published in 1948 in an article titled "A Mathematical Theory of Communication." Shannon was working on a classic engineering problem of signal transmission: how best to encode a signal, send it via some noisy medium, and decode it at a receiver. He conceived a way of calculating the *amount* of information in any given message and derived a formula to calculate the optimal rate of information transfer relative to the noise.

Shannon's deep insight—one that let him develop a theory of information in the abstract, independent of the message or medium—was to think of information in terms of uncertainty or probability. He reasoned that the way to calculate how much information it would take to send a message—a sequence or "string" of elements—is to ask how many other arrangements a string of the same length could be in.

If the string is just a single digit, say, then from the receiver's point of view, there are ten possibilities for which one it could be. If it were a single letter, then there would be twenty-six possibilities. The receiver has *greater uncertainty* about which letter it would be than which number it would be, because there are more possibilities. Conversely, the message with a particular letter contains *more information* than the one with a particular number, because it resolves a greater degree of uncertainty. Obviously, longer strings carry more information than shorter ones.

The amount of information can be measured in *bits*—roughly, how many yes/no questions you would have to ask to know what the message says. The amount of information in the signal thus relates to how *improbable* it is, which is a function of how many other forms it could have taken. If this sounds similar to the discussion in chapter two about the probability of particular arrangements of matter, it should—there is a mathematical equivalence between these ideas.

There is something very fundamental about this framing that is crucial for understanding the role of information in living systems. For something to count as information and for it to have causal power in a system *as information*, the possibility must exist that it could have been different. We will talk about this concept more in later chapters, but it's worth noting here that in a truly deterministic universe, such possibilities would not exist nor would information in this sense.

Shannon's ideas revolutionized signal transmission and laid the foundation for information technology and the digital age. But there's something paradoxical about the Shannon sense of information, which conflicts with the colloquial meaning of the word. In the Shannon sense, the information in a message does not have to be *about* anything. In fact, it takes more information to encode and transmit a completely random string of signals, like this—"akfh stwiol fbsdy vfln ud yriqpk"—than one like this: "this little piggy went to market." The first message cannot be compressed at all: if you want to transmit it accurately, you have to send every element. The second message, by contrast, has lots of statistical structure to it, based on the usage and common sequences of individual letters in English. If you know, for example, that a "th" is always followed by an "r," "w," or a vowel, then the information you need to send about the next letter is only one of eight possibilities, not one of twenty-six. So it requires less information to send the second message than the first.

But the first string—the one that has higher Shannon information—does not *tell you anything.* It takes information to send it, but the message does not carry information about anything else: if you know what the message is, all you have learned is what the message is. Shannon information just measures the coding complexity and redundancy of the message itself. It is very helpful for thinking about things like signal compression and channel fidelity and cryptography; in later chapters we discuss principles of signal transmission through noisy channels when we look at how information is transmitted in the nervous system. But Shannon information does not refer at all to what the message might mean.

As we will see, it is meaningful information—information *about things*—that living organisms need when trying to infer what is out in the world. *Meaning* is often treated in scientific discourse as something mysterious and difficult to quantify. But it rests on a commonplace and uncontroversial fact: some things are physically correlated with other things. Shannon called this "relative information," where knowing the physical arrangement of one thing tells you about the physical arrangement of something else.

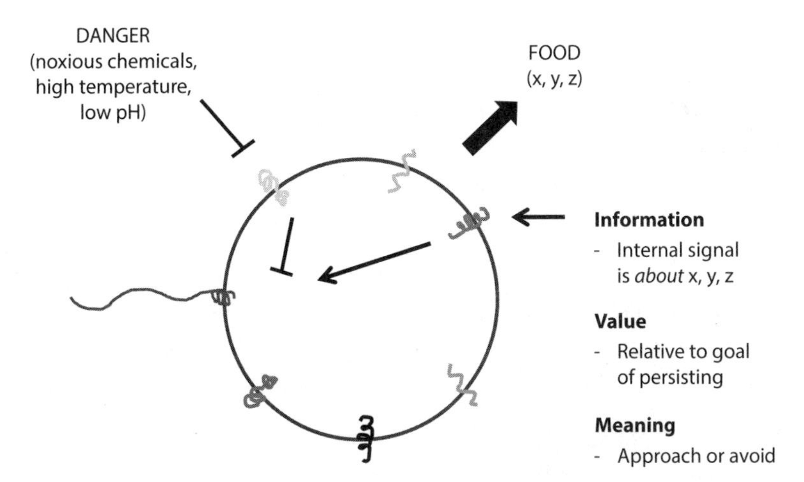


FIGURE 3.5. Information, value, and meaning. Unicellular organisms respond selectively to information about things in the outside world in ways that are dependent on their internal configuration. This configuration is shaped by natural selection to embody adaptive control policies: moving away from harmful substances and toward useful resources. Value (good/bad), relative to the purpose of persistence, and pragmatic meaning (approach/avoid) are thus configured into the physical organization of the organism.

Cells have relative information when the physical state inside them corresponds to or reflects something about the state of the environment (see Figure 3.5). Crucially, it is not due to a simple conveyance or transfer of physical forces. The cell's membrane constitutes a physical barrier, blocking the flow of energy and matter. Instead, for many receptor proteins, it is just a change in conformation of the bit inside the membrane that correlates with binding something outside. Even for receptors that signal by opening a channel for ions, it is not a transfer of energy or matter that is doing causal work but a change in the distribution of charge. It is relative information that is being transmitted.

However, moving from relative information to meaning crucially involves one additional element: the receiver. The symbols on this page do not have any meaning in themselves; they have to be interpreted by you, the reader. Similarly, it is not enough for a signal to be physically correlated with—to *signify*—something outside a cell. The meaning of the signal does not inhere in that signification alone but in *the act of interpretation*. In the organisms we've been discussing, that act of interpretation lies in how the system pragmatically responds to the signal: by a reconfiguration of internal dynamics or by approach or avoidance.

## Having Reasons

For a living organism, the meaning of a signal is rarely neutral but instead is intimately tied up with its relevance and value for survival. Before there were living organisms, there were no good things and bad things. Things are not good or bad in themselves: they only have meaning and value *with respect to* some goal or purpose. For living creatures, good things are ones that increase persistence. And bad things are ones that decrease persistence.

But note that the value of something out in the world is not inherent solely in the thing itself. It's not in the signal about the thing. It's not even in the action that ensues. It is instead in all those things, as viewed through the feedback from natural selection. Something is good if the tendency to move toward it is rewarded as an increase in persistence by the external critic that is natural selection. And something is bad if moving away from it is similarly rewarded (or moving toward it is punished). The value thus inheres in the whole system: the organism interacting with its environment and the consequences of these interactions over time.

The result of this feedback loop running relentlessly over millions of years is that organisms are configured to *do things for reasons*. If we want to understand their behavior, it is not enough to explain the physical mechanisms that transduce and integrate sensory signals and parlay them into action. That's the answer to how the behavior is mediated. But it does not capture the real range or even the most important type of causation in the system. We want to know *why* the organism behaves in that fashion.

"Why" questions are often seen as somehow unscientific, as if they betray an implicit anthropomorphism or the attribution of intentions to what are actually mechanical, inanimate processes. But "why" questions are really just "how" questions in a broader context and over longer timeframes. If we look at an organism and see that it tends to respond in a certain way to a given signal, we could say that the stimulus is the cause of that behavior. But this misses the larger point. The stimulus may be a trigger, but it is the particular configuration of the organism that *causes that signal to cause that behavior*. And that configuration is the outcome of eons of natural selection, which has pragmatically wired reasons for doing things into the structure of the living system. Evolution packs causal potential into life: like potential energy, this causal potential can be used to do work in the sense of directing the behavior of the organism.

These simple organisms are not aware of those reasons. But it is still correct to say that the organism is doing something *because* it increases its chances of persistence. Or, at a finer level, that it is moving in a certain direction *to get food* or *to escape a predator*. It's right to think of various components and subsystems as having functions. And it's right to say the organism is acting on the basis of inferences about what is out in the world, rather than simply being triggered by external stimuli. The mechanisms are simply the means by which those goals are accomplished.

Even these humble unicellular creatures thus have real autonomy and agency, as organized dynamic patterns of activity, causally insulated from their environment, and configured to maintain themselves through time. It is not merely that they hold themselves apart from the world outside: they act in the world and on the world in a goal-directed manner. They are causal agents in their own right.

As evolution proceeds, the degree of autonomy increases—at least along some lineages, like the ones leading to humans. The tight coupling of perception and action is loosened. With the advent of multicellularity and especially the invention of nervous systems, additional layers of processing emerge. Organisms evolve the means to represent sensory information internally without directly acting on it. More sophisticated control systems emerge for guiding action over longer timeframes. Organisms develop internal systems of evaluation that free them from the brutal, life-or-death judgment of natural selection. Crucially, all these systems are informational. Meaning becomes the currency of cognition.

# 4

# Life Gets Complicated

We now have the basic elements of agency in place: self-sustaining systems insulated from the environment that are working to keep their internal dynamics going and choosing from a repertoire of possible actions by (1) integrating information about internal and external states and (2) interpreting the meaning and value of that information relative to the goal of surviving. Even humble single-celled creatures are thus acting for reasons.

But they do live pretty basic lives. Their goals are closely aligned to the fundamental imperatives of survival. Eat and don't get eaten (or destroyed in some other way). Move toward sources of food and away from sources of danger. Reproduce when resources are plentiful; hunker down when they're not. At this stage of evolution, natural selection acting over generations ensures the adaptive behavior of individuals in the present. But the timeframes of decision making and purposive action are quite short. Primitive forms of learning can occur, but their effects only last for minutes or hours. Signals are integrated but are still tightly coupled to action. Meaning is not separately represented or apprehended—it is acted on. Single-celled creatures have a kind of mechanistic cognition, powerfully adapted to specific purposes but not yet reflective and not very flexible.

It is a far cry from the kind of sophisticated cognition and agency we see in more complex animals, which ultimately is elaborated to a level that might qualify as "free will" in humans. Exercising free will requires an open-ended ability for individuals to learn, to create new goals further and further removed from the ultimate imperatives of survival, to plan over longer timeframes, to simulate the outcomes of possible actions and internally evaluate them before acting, to decouple cognition from action, and ultimately to inspect their own reasons and subject them to metacognitive scrutiny.

But we are on our way. The primitive building blocks in unicellular creatures provide the substrate for evolution to create more complex systems with ever-greater autonomy. This may sound like an inexorable, even a purposeful process, as if greater complexity were a goal of evolution. It is not. Evolution has no purpose. Whatever works will be retained. Simple organisms have persisted for billions of years because the strategies they pursue are very well suited to their environments. But there can be a directionality toward greater and greater complexity along certain lineages. This is because each step explores new niches not previously accessible and, in the process, opens up yet more new territory in ecological space to be explored and exploited. Like the evolution of the digital economy that we are witnessing in our lifetimes, each increase in complexity makes subsequent increases both possible and profitable.

## The Energy Barrier

Yet there is a barrier—an energetic one—to be overcome before gaining access to this space of increasing complexity. We saw how ruthlessly streamlined bacterial metabolism must be. Bacteria only make proteins when they need them—like those enzymes for digesting lactose, for example. Indeed, if there is no recurrent need for some specific proteins, the genes themselves are often lost from the bacterial genomes simply because it takes too much energy and time and raw materials to copy them during cellular reproduction. Genes and proteins are costly, and for bacteria, energy is in limiting supply.

Bacteria generate ATP—the cell's internal energy currency—by allowing hydrogen ions from outside to rush through the channel in the ATP synthase proteins that sit in their outer membrane. But there's a limit to how many such proteins they can fit in their membrane, especially because they need room for the protein that pumps hydrogen ions back out of the cell to generate the needed gradient in concentration between the outside and the inside (as well as other things like sensory receptors and flagella). Given these physical limits, bacteria just cannot generate enough ATP to power a truly complex cellular economy.

For life to become bigger and more complex, this energetic hurdle had to be leaped—and it really was a leap. Something drastic had to happen beyond the normal, gradual evolutionary processes of "descent with modification," where each generation differs only in tiny ways from the one before. Little tweaks were never going to (or at least never actually did) enable bacteria to go beyond this limit. What did the job was something else entirely: a symbiosis.

Life is divided into two major domains: (1) prokaryotes like bacteria and (2) eukaryotes, which include plants, animals, and fungi. Prokaryotic cells are small and relatively simple. They evolved first and existed for more than a billion years before the first eukaryotic life arose on Earth. Eukaryotic cells tend to be much larger and have a far more complex cellular structure characterized by a nucleus where the DNA is enclosed in an internal membrane, as well as multiple types of specialized internal compartments called organelles. These organelles include mitochondria—subcellular structures, each bounded by its own membrane, that are the site of ATP production in eukaryotic cells.

Mitochondria are curious little things—they are the inspiration for "midichlorians" that are supposed to generate "The Force" in *Star Wars*. They are often called the powerhouse of the cell, and they do indeed function like little power plants, pumping hydrogen ions out into the cytoplasm of the cell and using the resultant gradient to power the ATP synthase protein that spans their membranes. But they're more than just a functional compartment of the larger cell. They also have their own private set of genes, which encode proteins required for the biochemical processes that happen inside the mitochondrion. Mitochondria and their genomes also are replicated independently from the genes in the nucleus and separately partitioned into daughter cells during cell division.

Indeed, in looking at that biochemistry and at the overall structure and life cycle of mitochondria, it was proposed as early as 1905 that mitochondria are in fact the remnants of bacteria that were somehow engulfed by a larger cell and that became "endosymbionts": a life-form living inside another life-form. That remarkable hypothesis was revised and greatly expanded by evolutionary biologist Lynn Margulis in the late 1960s, and though it was considered outlandish at the time, it is now supported by overwhelming evidence.

The host cell for this endosymbiosis event is likely to have been a member of another domain of prokaryotes, known as archaea, that are quite distinct from the types of bacteria we discussed (see Figure 4.1). Archaea were first discovered living in extreme, hostile environments like hot springs or salt marshes and eating (or "breathing") things like methane and sulfur. Members of this group have since been found in all kinds of more forgiving environments, even in our own gut, for example.

Archaea are like bacteria in many ways, but they also have important differences in their biochemistry, especially in the proteins they use to make DNA and RNA. These look much more like the ones that do these jobs in eukaryotic cells, which provides evidence that some ancient archaeon acted as a host for an endosymbiotic bacterium, thereby giving rise to eukaryotes. There are a variety of theories for how this may have happened, but they all involve a big archaeon engulfing but not digesting a smaller bacterium. Instead, the two life-forms established a mutually beneficial relationship, with the bacterium surviving in the relatively sheltered and supportive environment of the archaeal cell while providing its host with an internal resource to generate additional energy.

Over subsequent evolution, many of the genes inside the bacterium were lost because they were no longer required or their functions were transferred to genes in the nucleus of the host cell. The bacterial genome shrank to the bare minimum required to sustain its own special biochemistry, which was now supplying ATP to the whole cell. Thus, the bacterium could no longer survive on its own, but neither could the host: they became entirely interdependent and established a new branch of life. (A similar endosymbiotic event of bacteria capable of photosynthesis subsequently led to the development of chloroplasts in plants.)

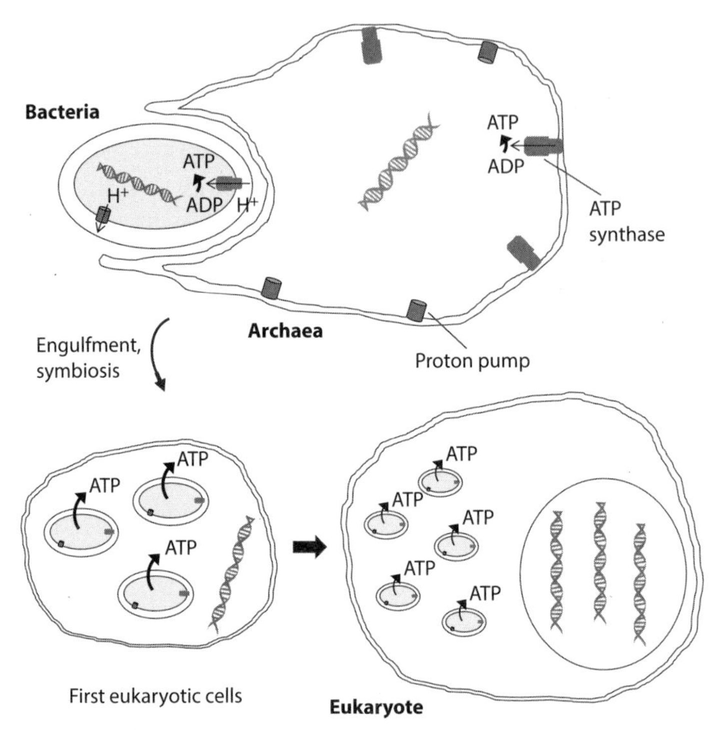


FIGURE 4.1. Endosymbiosis. The first eukaryotic cells are thought to have arisen when some ancient archaeal single-celled organism engulfed a bacterium, giving rise to a symbiotic relationship. Over time, many of the endosymbiotic bacterial genes were lost or transferred to the nucleus of what became a eukaryotic cell, and the bacteria became mitochondria—sheltered within the larger cell and generating energy for it. The surplus energy resulting from this living arrangement allowed eukaryotic cells to maintain more genes and increase in size and complexity.

The impact of this evolutionary event was similar to that of the invention of agriculture in human prehistory. Humans had scraped together a living as hunter-gatherers in small bands for hundreds of thousands of years at least before plants and animals were domesticated. This was a precarious existence, one that did not give much time for anything but finding food. The agricultural revolution brought an increase in population size and density, growing specialization in diverse new types of labor, a need for more complex systems of governance, the development of new tools and technologies, with surpluses of food and other goods driving the emergence of industry, commerce, and trade with other communities. In short, it enabled and increasingly drove the emergence of civilizations supported by diverse and complex social systems.

The domestication of bacteria and their gradual transformation into mitochondria was a similarly revolutionary process. It gave archaeal cells (now the first eukaryotes) not only a ready supply of energy but an energy surplus, enough to let them escape the need to stay lean and mean. They could get bigger—much bigger—but more importantly, they no longer had to be so frugal, especially regarding the maintenance of genes. They could hang on to more genes, some of which could take on new roles, leading to a greater scope for new cellular functions and structures and a more complex repertoire of possible cellular states.

Eukaryotic cells could thus react more flexibly to changing environments, allowing them to diversify into new niches. Crucially, they could effect a division of labor in cellular cooperatives to generate the first truly multicellular organisms. With added levels of internal processing and an expanded repertoire of possible actions, multicellularity brought ever-greater autonomy and agency—the power of the organism itself to choose. But this increasing sophistication of behavior required a correspondingly sophisticated system to control it.

## E pluribus unum

We already saw transient multicellular behavior in the slugs and fruiting bodies formed by the aggregation of individual Dictyostelium amoebas. This kind of aggregative multicellularity is observed in many other species, across diverse groups of eukaryotes, and even in some bacteria called myxobacteria. It usually involves the differentiation of a small number of cell types that play different roles in the collective. Such aggregation and coordinated differentiation confer a benefit to the individuals engaging in it, allowing them to persist through harsh environmental conditions or food shortages by forming spores, which get dispersed from the fruiting bodies.

However, this is only a benefit to *some of* the individuals who make up the aggregate; namely, those that become or make the spores. The success of the whole multicellular aggregate relies on a division of labor; some individual cells (up to that point free-living individual organisms) have to forgo the opportunity to reproduce and instead form the stalk so that others can successfully reproduce. So why would they do it? Why would evolution select for that behavior?

As we saw, evolution selects for behaviors that increase the persistence of the dynamic pattern of biochemical processes of each individual. When a genetic template provides the instructions for that pattern, persistence can be achieved or, even better, turned into amplification by reproduction and recapitulation of the pattern in a new cell. However, within a population of unicellular organisms, individuals are usually in competition with each other for limited resources. Any genetic variation that gives an individual an advantage in persistence or reproduction increases the frequency of that particular genetic variant *at the expense of others*. This doesn't sound like a setup ripe for the kind of social cooperation evident in multicellular aggregation.

In fact, it sounds like one that would foster competition within a multicellular collective of Dictyostelium amoebas to contribute more to the subpopulation that forms spores and thus reproduces. This is exactly what happens—or at least what would happen if countermeasures had not also evolved. If cells with different genomes aggregate, there will be evolutionary pressure to contribute disproportionately to the spores, and a genome that tends to promote that behavior would increase in frequency in the population. In turn, this would create a countervailing pressure to prevent such "cheating," so that genomes that evolve that ability will then do better. An arms race of cheating and cheating detection and detection evasion then arises—a dynamic observed in all kinds of social situations where a balance between cooperation and competition is at play.

The easiest way to avoid this escalating conflict is to ensure that only cells with the same genome aggregate. If all the cells are genetically identical (because they were all generated by clonal division from the same founding individual), then there is no conflict. From the point of view of evolution, the important thing is that the genome gets copied. Cells that contribute to the stalk thus share the same goal as those that contribute to the spores. They can ensure persistence of their "own" pattern by indirect reproduction; that is, the reproduction of other cells that share the same pattern because they share the same genome.

It is thus not surprising that mechanisms evolved in many of these species to ensure that only genetically identical cells can aggregate together. In Dictyostelium, for example, a pair of proteins that span the cell membrane mediates the adhesion of individual cells. The genes that encode these proteins are highly variable, however, so that, in a population, different individuals express slightly different versions of the proteins. Only cells expressing the same versions aggregate together. If two clonal populations are initially mixed, with individual cells of both types dispersed, then under conditions that favor aggregation, these two populations will largely segregate into two separate aggregates. Analogous systems exist in other aggregating organisms, using different sets of proteins, with this convergent evolution in diverse lineages illustrating the importance of having such a mechanism.

These individual organisms and the resulting aggregates are thus capable of distinguishing "self"—has the same genome as me—from "non-self." This property is not absolutely essential for aggregation to occur; many organisms form collectives in which all the individuals derive some benefit, such as flocks of birds or schools of fish, for example, or even the complex ecologies of bacterial biofilms. The benefits may be shared, but they are manifested at the level of the fitness of each individual. Something very different happens when the individual organisms contributing to the collective are all clonally related. They share a common goal and happily divide labor to achieve it.

Like the Borg in *Star Trek*, these individuals get assimilated into a collective. It is not just that resistance is futile: it's pointless. Why should I resist joining forces with more me? When genetically identical cells come together in this way, they form a truly new individual—a new, unified self at a higher level of organization. The fitness of all the individual cells gets transferred to the fitness of the whole multicellular organism. Some cells specialize in reproduction, others in states or functions that favor survival: growth, motility, defense, and so on. Anything that benefits the fitness *of the whole organism* is selected for, even if, considered in isolation, it comes at the expense of the fitness of individual cells.

In contrast to isolated unicellular organisms that adjust their internal physiology and select actions to further their own persistence, these processes now occur at the level of the whole multicellular collective. This new kind of organism thus becomes a locus not only of fitness but also of agency.

## Big Hungry Beasts

Coming together is one way to achieve multicellularity. But an even simpler way—and one that ensures genetic identity—is *staying together*. If cells replicate but do not separate, remaining attached to each other instead, then a multicellular organism can arise. This is, of course, exactly what happens in most species of animals, which start from a single cell—the fertilized egg or *zygote*. This cell divides to make an embryo with two and then four and then eight and eventually thousands or millions or billions of cells.

The multicellular lifestyle has many benefits. First, organizing into a big mass provides shelter for the cells on the inside from the vagaries of the environment, with cells on the outside often specializing to form a protective barrier. As in aggregating unicellular organisms, a small set of cells—the germ line—become dedicated to reproduction, with all the others forming the *soma* or body dedicated to ensuring that the organism survives long enough to reproduce.

Second, getting bigger makes it easier to eat littler things and harder to be eaten. As multicellularity evolved, the resultant dynamic of predator and prey would have created all kinds of new ecological pressures and imperatives for perception and action. If what you are trying to eat is doing its best to avoid being eaten, and if you must similarly do your best to avoid becoming something else's lunch while catching a meal yourself, then entire arsenals of predation, as well as defense and counterdefense, would emerge.

These kinds of specializations are costly, of course, but the economies of scale conferred by multicellularity further increase the energetic surplus provided by the acquisition of mitochondria. This surplus, in turn, gives greater scope to increase genetic complexity, and having more genes enables the differentiation of more diverse cell types, tissues, organs, limbs, sensory structures, and so on. This opens up whole new lifestyles and niches, as well as new opportunities for diversification and elaboration, with each evolutionary move granting access to yet more possibilities and creating pressures to exploit them.

Multicellular organisms face all the same challenges as unicellular ones in keeping themselves organized, maintaining their vital dynamics, and reproducing. To perform these functions effectively, they have to be able to sense relevant things out in the world and respond appropriately by altering their physiology or their behavior. But a multicellular creature faces an added challenge: those physiological or behavioral responses require coordination of individual cells across the whole body. And for behavioral responses, which may involve moving in the world or moving just a part of the body—one of those specialized limbs, for example—the pressures of predation also mean those actions need to be made quickly!

Achieving that kind of control required the invention of new types of cells and tissues: muscles and neurons. To see when—and why—these innovations arose, we need to look back into evolutionary history, deep into the roots of our family tree, to infer what kinds of creatures existed at different stages.

## La Familia

We have two sources of information available to us when making inferences about our ancestors. First is the fossil record, which gives an account of the life-forms extant during different geological periods. This record is hugely useful, tracking the emergence of new kinds of creatures over the course of evolution. But it is also very incomplete.

It captures creatures with hard parts—shells and skeletons—much better than soft-bodied, squishy animals, whose bodies tend not to fossilize so well. Thus, it gives a biased view of the abundance of certain kinds of animals, which is based on how conducive their environments were to the processes of fossilization.

There is another record that we can turn to, however, which is the DNA of creatures that are now alive. Just as genomics databases can help you track down distant relatives, and tell you whether they are first or second or third cousins by comparing your DNA to theirs, the sequences of all creatures carry the same record of their relatedness, over vastly greater timespans. Humans, for example, are most closely related to chimps and, after that, to gorillas, then orangutans, then other apes, then monkeys, and so. That does not mean that we descended from chimps—it means that chimps are our cousins and we share a common ancestor with them.

That common ancestor was probably a lot more like a chimp than a human. We know that because we don't see anything in the fossil record that looks like a human until quite recently. So, if we put the evidence of genetic relatedness together with the fossil record, we can make some inferences about what distant ancestors of current creatures looked like at various branch points, where species diverged and new forms emerged. When those divergences happened, sometimes one of the species remained quite like the ancestor in form and habits, continuing to successfully exploit its niche, while the other evolved along a novel pathway into a new ecological niche. (Indeed, the movement of some individuals into a new niche is often what caused the divergence.)

If we do those comparisons across all existing animals and calibrate them with the fossil record, we can build an evolutionary tree showing the emergence of new forms. Although it is incorrect to think of evolution as a march of progress, because simpler forms continued to thrive in their own niches, it is nevertheless possible to track innovations over time that led along certain lineages to increasing complexity and, more importantly for our purposes, to more sophisticated kinds of agency. To get a picture of how the ability to act in the world evolved in animals, we can start by tracking the ability to coordinate bodily movement.

## Coordinating Movement with Electricity

There's no point having a great big body with loads of cells if they all still act like independent organisms, and especially if they try to move as individuals—that could cause the body to literally pull itself apart. To be able to act on the world, the multicellular organism has to move as a unit or be able to move parts of itself in a coordinated fashion. We saw, in simple creatures like amoebas (individual Dictyostelium, for example), that movement relies on specialized proteins that make up an internal cytoskeleton—long fibers that can be moved with respect to each other, generating both a fulcrum for movement and a propulsive force. In the evolution of animals, the same types of proteins got pressed into service in a new cell that was specialized for movement: the muscle cell.

In muscles, these cytoskeletal proteins are aligned parallel to each other within each cell, and many individual muscle cells align in parallel to make large muscle fibers. These fibers can contract to generate a pulling force in the direction of alignment of all the fibers, which can be strong enough to move the rest of the body. But how to get them to contract in unison? This process relies on a signal, one that we saw before in Paramecium: an electrical impulse that opens ion channels in the membranes of the muscle cells, allowing in this case an influx of calcium ions, which triggers the movements of the cytoskeletal proteins over each other, resulting in contraction of the whole muscle.

That's all fine, but it's not enough. To move its body effectively, an animal has to be able to selectively contract some muscles while relaxing others (just as extending your arm requires contracting your triceps while relaxing your biceps). This kind of coordination requires rapid communication over distance. A lot of signaling between cells is mediated by the release of proteins or other molecules and their detection by receptors on the surface of other cells. This kind of chemical signaling is cheap, energetically speaking, and very effective for some things. But it is slow, relying on the passive diffusion of molecules, and it's hard to direct to specific sites or target to specific cells. Multicellular organisms still do use this kind of signaling, especially for regulating physiological processes on a scale of minutes to hours. But the coordination of behavior requires something more targeted and much, much faster—on the scale of milliseconds. That is where neurons come in.

Neurons are highly specialized for the conduction of fast electrical signals to specific targets over potentially long distances. They didn't invent this kind of bioelectrical signaling; it relies on components that we already discussed: ion channels and pumps. These proteins allow cells to manage the concentration of ions, such as sodium, potassium, calcium, and chloride, on their inside relative to the outside (thus regulating the *osmotic pressure* the cell is under—how much water wants to move in or out of it). At the same time, they allow cells to manage the distribution of electrical charge that these ions carry.

The resulting concentration gradients of charged ions generate an electrical potential. This creates an opportunity: opening ion channels in the membrane in response to some external stimulus lets charged ions flow down that potential gradient and act as an internal signal to trigger some adaptive cellular response. Many ion channels act as sensors or receptors in that way for different sorts of external stimuli, such as chemicals, light, mechanical displacement, and so on (see Figure 4.2). That is one of the ways that unicellular creatures sense things in their environment, and it's the first step in many of our own senses.

But there is another type of ion channel that is triggered by a change in voltage across the membrane. Such channels can be used to rapidly amplify and shape a much bigger and more global electrical response of the cell to initial stimulation of one of the receptor-type channels. Once some ions flow into the cell, they change the voltage across the membrane (which depends on the relative concentration of ions on either side). This can cause some voltage-gated channels to open, which further changes the voltage, which opens more channels, and so on. This can lead to an electrical pulse of the type we saw in Paramecium: a fast, global response to a sudden stimulus.

If you're wondering why the cell does not just explode when this happens, it is because the force driving the pulse is rapidly counteracted. First, the change in concentration weakens the electrical potential driving the ions into the cell. Second, the ion channels have an intrinsic

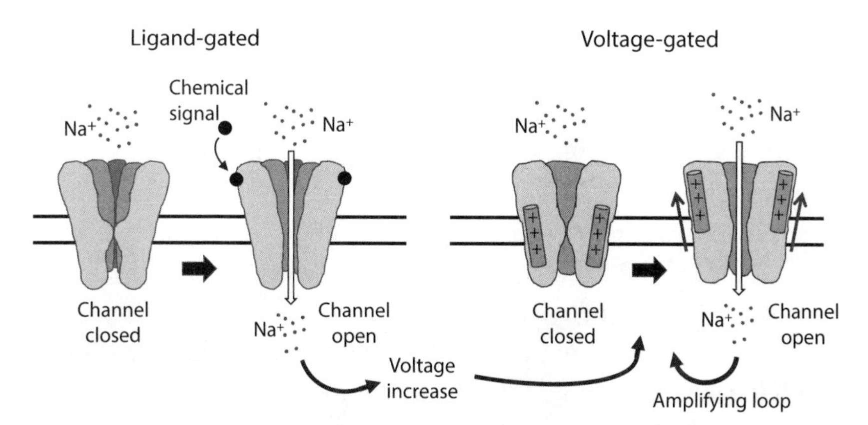


FIGURE 4.2. Ion channels. Ion channels are typically multiprotein complexes that span the cell membrane, forming a pore through which certain chemical ions can pass, such as sodium ($Na^+$), potassium, or chloride. The opening and closing of this pore can be regulated by binding to a particular chemical like a neurotransmitter (ligand-gated, left) or in response to voltage changes across the membrane (voltage-gated, right). The combined action of these types of ion channels is responsible for synaptic transmission and the generation of electrical action potentials or *spikes*.

mechanism that shuts them very quickly followed by a *refractory period* during which they cannot open again. So, there are some passive mechanisms that tend to make the pulse peter out. But more importantly, other ion channels are responsive to the change in voltage that occurs during this pulse and begin to get recruited and conduct other ions in the opposite direction. This actively restores the *resting potential* of the cell—the normal concentration of different ions on either side of the membrane—and means the pulse ends almost as quickly as it begins.

Lots of different types of cells make use of electrical signals, either in these kinds of rapid pulses or in a more gentle, graded fashion to trigger the contraction of muscle fibers or the secretion of hormones or other chemicals. Many can even conduct or transmit electrical signals directly to neighboring cells through special channels called *gap junctions*. But neurons are the real masters when it comes to harnessing this mechanism.

## Neurons

Neurons are highly specialized for the generation and conduction of electrical signals, but they communicate with each other through the release and detection of special chemicals known as *neurotransmitters* at specialized structures called *synapses*. When an upstream neuron releases a neurotransmitter, it is detected by receptor proteins on the synaptic membrane of the downstream neuron. Some of these receptor proteins are ion channels that open in response to this signal. If enough ions enter the neuron within a short enough timeframe (integrated across all its synapses), this initiates an electrical signal as described earlier. (Other types of receptor proteins initiate chemical signaling inside the synapse that modulates the threshold for electrical activity.) When an electrical pulse is sent down the axon, it in turn induces release of the neurotransmitter onto downstream neurons.

However, the real novelty of neurons is not the use of electrical signals within each neuron or the transmission of chemical signals between them. It is their morphology that makes them special and so well adapted to their roles in long-range communication and coordination and in the processing of information.

Cells come in all shapes and sizes, but whether they are globular or packed together like little bricks, most are content to keep their bits local. Neurons, by contrast, send long, thin cellular projections out from their cell body, which branch to form wondrous treelike structures that enable them to connect with many other neurons or with sensory and muscle cells. Importantly, they can bypass many cells along the way, specifically connecting over long distances—up to a meter in the human body!

There are two types of these extensions: *dendrites*, which are specialized for receiving signals, and the *axon*, which is specialized for sending signals. The neuron is thus polarized: it has an "input" end and an "output" end, with the cell body, where the nucleus is, sitting in the middle. Indeed, it's tempting to think of neurons as input/output devices, analogous to components of electronic circuits, that can take in a signal, perform some kind of operation or transformation of it, and output a different signal.

Neurons do indeed do that; for example, taking a signal from a sensory cell and transmitting it to a muscle cell. But they are far more than simple linear connectors.

The real power of neurons comes from the way they are connected. They rarely have input from only one cell that they send to just one other cell: frankly, that would be pointless. Instead, their branching dendrites collect signals from many cells, allowing the neuron to perform all kinds of integrative operations to extract relevant meaning from that incoming information (see Figure 4.3). Similarly, they can send output to many cells, allowing them to convey information in a coordinated fashion across a network of interconnected cells. We will see in the next chapter how these activities in neuronal networks come to support sophisticated perception and cognition.

Yet, the first functions of neurons in an evolutionary sense may have been simply to coordinate the activity of fields of muscle cells. If we look across our evolutionary family tree, we find that our most distant relatives among multicellular animals are simple marine sponges. Sponges have a pretty boring and undemanding lifestyle. They are simple filter-feeders: they attach to some surface and filter food from sea water that they pull through their cuplike bodies by waving flagella on the cells lining their internal cavity. Some of their cells are capable of a simple kind of contraction, which they can coordinate by direct electrical coupling between neighboring cells. But they do not have any muscles nor any neurons—they had not been invented yet at the point at which sponges diverged from other animals.

But when we move a couple more branches along this early tree we come to creatures like corals, sea anemones, and jellyfish, which do have muscles and do have neurons. This group is called the Cnidaria (with a silent *c*), after the Greek word *cnida*, which means nettle: these creatures all have specialized stinging cells that they use for predation or defense. They are capable of coordinated movement either by moving through the water, like jellyfish, or moving parts of their bodies, like corals, which can reach out fleshy tentacles in a coordinated fashion. One such creature, the Hydra, provides an illuminating example of how a multicellular organism can coordinate its behavior.

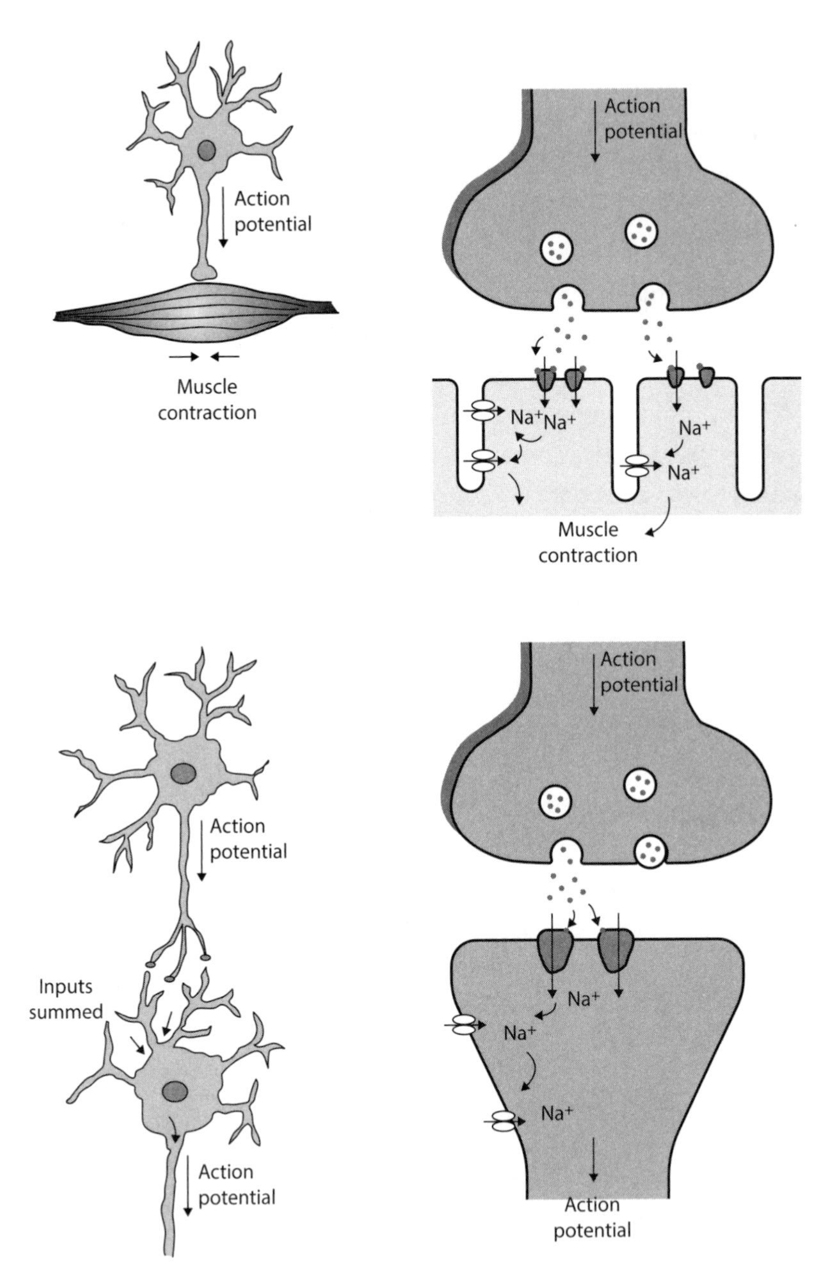


FIGURE 4.3. Synaptic transmission. When an action potential travels down a nerve fiber to the synaptic terminal, it results in the release of calcium ions, which increases the probability of the fusion of vesicles with the outer cell membrane, thus releasing neurotransmitter molecules. These molecules are detected by ion channels on the membrane of the postsynaptic muscle **(top)** or neuron (**bottom**), resulting in the increased probability of opening of the channels and the influx of sodium ions ($Na^+$). This signal is amplified by the opening of voltage-gated channels, resulting in the contraction of the muscle or the firing of an action potential in the postsynaptic neuron.

## Hail Hydra!

A simple little freshwater polyp, a Hydra is shaped like a fat, hollow cigar, about a centimeter long, with a ring of tentacles at one end surrounding its mouth and a footpad at the other end. What they don't have is an anus—they had not yet been invented. Food and other particles go in the mouth, and anything that's undigestible gets expelled by the same route. Hydras got their name from the fact that, if cut into two, each of the segments can regenerate a whole new body, and some Hydras can even grow multiple heads like the creature from Greek mythology. They spend a lot of time adhering to rocks or vegetation with their footpad, but they are far from inactive: they use their tentacles to feed or defend themselves, make a variety of bodily contractions and extensions, and can also detach and move around in a purposeful fashion.

Although Hydras have neurons, they do not have brains—they also had not yet been invented. Their neurons are organized in a diffuse *nerve net*, which is laid out in contact with the two layers of cells that make up the body of the animal: one lining the internal digestive cavity and another on the outside of the animal (see Figure 4.4). Both layers of cells are somewhat similar to our muscles: they are capable of contraction in response to electrical stimulation but they also multitask. The ones on the inside also have digestive functions, and the ones on the outside also function as protective and externally sensitive skin.

The behavior of the Hydra has been studied for more than three hundred years, but it's only recently that some of the underlying principles of its neuronal control were discovered. Pioneering work in the lab of Rafael Yuste at Columbia University has documented the complete behavioral repertoire of the Hydra while simultaneously imaging the activity of all the muscles and all the neurons of the entire animal.

It turns out that Hydras are capable of only a small number of basic movements: longitudinal contraction, longitudinal extension, radial contraction, bending, releasing their footpad, extending their tentacles, retracting their tentacles, and opening and closing their mouth. The bodily movements are coordinated through the activity of specific subsets of neurons. Although the nerve net looks quite uniform, recent

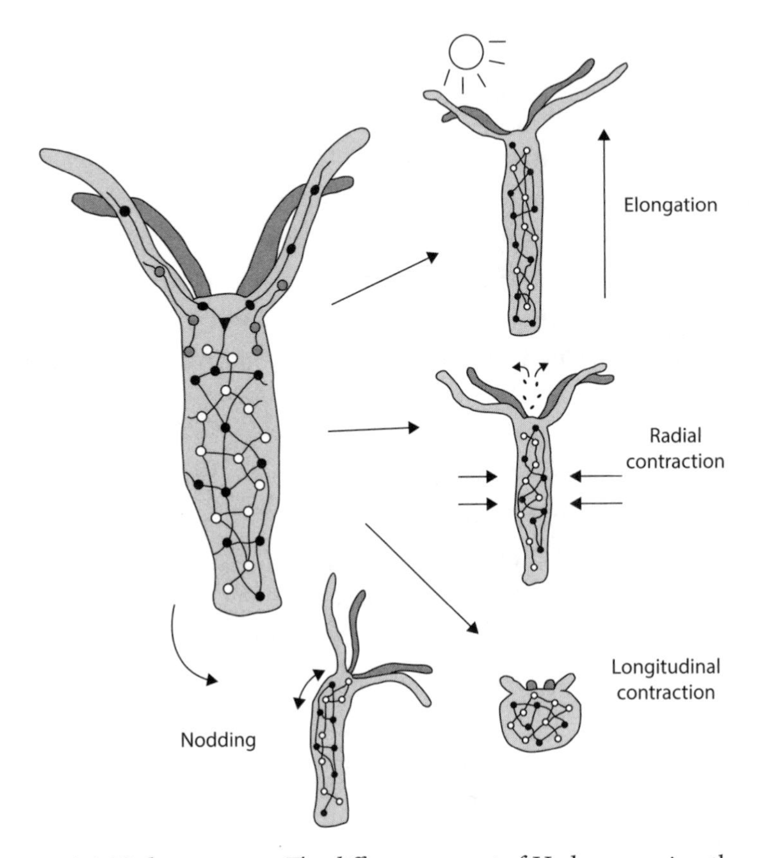


FIGURE 4.4. Hydra nerve net. The diffuse nerve net of Hydra comprises three or four intercalated subnetworks that control the animal's small repertoire of actions.

imaging work and molecular characterization reveal three distinct subnetworks, intercalated with each other and innervating different body wall muscle cells, as well as a smaller network at the base of the tentacles around the mouth. When one of these distinct networks activates, it triggers one or another of the bodily movements while inhibiting other movements.

These basic movements can be executed individually or deployed in more or less fixed patterns or sequences to execute different behaviors. For example, in feeding, the Hydra extends its tentacles outward, and when one of them touches some object, this triggers the explosive release of a little barbed harpoon from the stinging cells on the tentacles. If this harpoon pierces something, tension on the barbs causes a release of toxins that paralyze the prey (if the object is indeed prey). The Hydra then pulls its tentacles back, opens its mouth, and ingests the hapless victim.

Locomotion also involves stereotyped sequences of various kinds of bodily movements. In one mode, the Hydra bends over, grabs with its tentacles the surface it has been "standing" on, releases its footpad, then bends double again, reattaches its footpad in a new position, and releases its tentacles. It can keep this "somersaulting" up for several iterations (though I think "cartwheeling" is a more apt term for it). Alternatively, it can bend over, release its footpad, and simply walk upside-down on its tentacles for some distance. Finally, it can detach from the surface altogether, blow a little air bubble, and float away on whatever currents are available.

Note that the behaviors here are rightly described as hunting or feeding or moving from one place to another. Those words describe *what the organism is doing*. The set of movements it employs—contracting, bending, detaching, and retracting various parts—is merely how it accomplishes those goals. The structure and wiring of the nervous system and the musculature define a repertoire of possible movements of the body and allow the organism to coordinate the deployment of these bodily movements to execute goal-directed behaviors.

Of course, there's little point to being able to choose among different behaviors if you do not have any information on which to base such a choice. Another crucial function of the nervous system is to allow action selection to be informed by external or internal signals that reflect either what is out in the world or the internal state of the organism. In the Hydra, responses to external signals include moving toward light; responding to mechanical touch; regulating whether to actually ingest something that the tentacles have snared, which depends on the detection of a chemical only secreted by (wounded) living things; moving away from waters at unhealthy temperatures; and so on. The frequency of various behaviors also varies depending on how well fed the Hydra is, whether it has been in the dark for some time, and other parameters reflecting its recent experience and current state.

It is important to emphasize that the organism is not just sitting passively, waiting for some stimulus. Hydras engage in spontaneous behaviors, and their nervous system shows endogenous patterns and oscillations of activity, even in the absence of any external stimuli. A Hydra is not just a passive stimulus–response machine. It is an inherently, incessantly dynamic system and an autonomous agent—one that accommodates to incoming signals and uses them to inform its ongoing selection of possible behaviors.

That said, it lives a simple life and doesn't seem (as far as people have been able to tell) to take advantage of the other amazing capacity that nervous systems bring: the ability to learn. Or at least, given its rather simple perceptual capabilities, it doesn't have much to learn about. The real expression of that capacity required some further evolutionary elaboration, exemplified by the next creature in our story, the tiny nematode worm, *Caenorhabditis elegans*.

## As the Worm Turns

The humble *C. elegans* is a much-loved model organism for genetics and neuroscience and has easily the best-characterized nervous system of any creature to date. These little worms are much smaller than Hydras: they are microscopic, in fact, at one millimeter in length. But even though their repertoire of movements is similarly limited, their nervous systems have a more complex architecture that supports more sophisticated "cognitive" operations and behaviors.

*C. elegans* have two sexes, males and hermaphrodites (which make both eggs and sperm and can self-fertilize, if there are no eligible bachelors around). They crawl around in the soil, moving in a snakelike sinusoidal fashion, looking for food (like poor little *E. coli*, which seem to get eaten by everyone) or for mates, while simultaneously avoiding an ever-changing array of threats. They are thus in a constant search for optimal conditions, balancing exploitation and exploration, managing their social lives, and reacting to changing circumstances that in turn often reflect their own actions.

These nematodes are remarkable for how stereotyped their developmental program is, resulting in bodies with exactly the same number of cells, of exactly the same types, laid out in exactly the same positions across individuals. Adult hermaphrodites have 959 somatic cells (i.e., not counting cells in the germline), and adult males have 1,033 cells. Of those cells, 302 in hermaphrodites and 387 in males are neurons (the extra ones in males mainly being devoted to control of their reproductive organs).

Compared to the Hydra's simple nerve net, *C. elegans* have a greater diversity of neuronal subtypes that are laid out in a more hierarchical fashion (see Figure 4.5). These subtypes include sensory neurons, responsive to touch or to chemicals; motor neurons, which connect to and activate muscles; and *interneurons* in the middle, which, as the name suggests, connect one neuron to another. Interneurons can be either excitatory or inhibitory; that is, when they send a signal to another neuron, it makes the target neuron either more or less likely to become electrically active, respectively. This enables the integration of different kinds of signals, with the excitatory–inhibitory balance of many inputs determining the electrical activity of the downstream neuron. It also provides the means for control by negative feedback and cross-inhibition.

As in the Hydra, the basic layout of neurons and muscles defines the repertoire of bodily motions available to the worm. For example, their sinusoidal pattern of movement is mediated by motor neurons spaced out along the body, innervating muscles on either side. When a motor neuron on one side is activated, an interneuron ensures that the matching motor neuron on the other side is inhibited, so that muscles on one side can contract while those on the other relax. Then a sequence of neuronal activation passes down the length of the worm to propagate the wave of muscle contraction that underlies the sinusoidal movement. A different set of neurons controls movement in the opposite direction, when the worm needs to back up.

The execution of one action at a time is coordinated by specific *command neurons* that, when activated, initiate one of the possible movements. Importantly, once the command neurons activate, the motor sequence (e.g., for moving forward) happens by itself. It does not need to be centrally directed because it is already wired into the neuromuscular circuitry: it simply has to be released (while all other possible actions are inhibited). The worm's brain (a condensation of interneurons

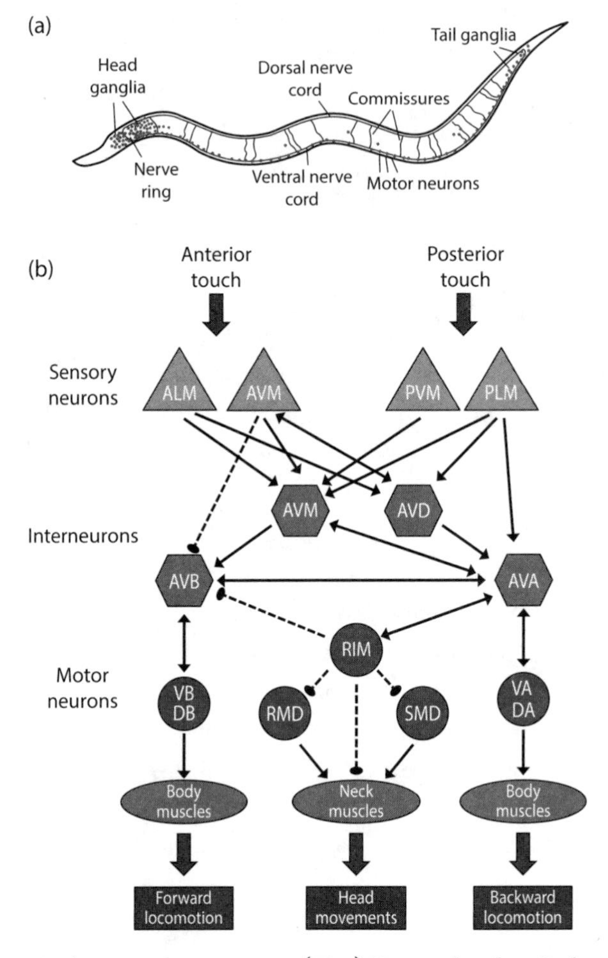


FIGURE 4.5. *C. elegans* nervous system. **(Top)** Hermaphrodite *C. elegans* have precisely 302 neurons, including sensory neurons of many types, motor neurons controlling muscles, and many interneurons concentrated in what we may call a "brain." **(Bottom)** Signals from diverse sensory neurons are integrated by a network of interneurons, which ultimately regulate command neurons to drive one behavior while inhibiting all the other options.

in the head) just has to ask the motor system to do something; it doesn't have to tell it how.

But it does have to decide which action to take, based on what it encounters in the world and its own internal state. Like many organisms, *C. elegans* innately find some chemicals attractive, such as those associated with its food sources, and some aversive, including various toxic chemicals or signals of unfavorable conditions. Accordingly, they approach or avoid such molecules, using strategies similar to bacteria to move up or down a perceived concentration gradient.

As mentioned, the valence of these signals—whether they are positive or negative from the point of view of the animal—is innately configured into their neural circuitry. Sensory neurons that detect a variety of toxic or noxious chemicals (or aversive stimuli like a bump on the nose) are wired to command neurons that cause the animal to reverse direction and go somewhere else. Conversely, neurons that detect chemicals associated with food or potential mates are wired to command neurons that promote approach. Note that the meaning and value of these signals, for the worm, are still pragmatic, tied up with its response and not yet independently represented. It is not that the worm moves away from something because *it senses that* it's aversive; it's aversive because the animal's ancestors moved away from it, and natural selection said that was a good idea.

But the coupling between perception and action is at least loosened a bit. There are now some intermediate stages of processing—carried out by the middle layers of interneurons—during which multiple signals are integrated to allow the animal to respond to the situation as a whole, as opposed to independent stimuli. Specific interneurons collect signals from multiple sensory neurons responding to diverse aversive stimuli, while other interneurons sum the activity of a different set of sensory neurons responsive to diverse attractive stimuli. The relative activity of these interneurons is then itself integrated at another stage to determine whether the sum of attraction outweighs the sum of aversion. All of this is dependent on the context: responses to those integrated external sensory signals differ depending on the current internal state of the animal.

Various parameters of internal state are signaled by specialized neurons that mainly use chemicals such as serotonin or dopamine (derived from simple amino acids) to communicate to other neurons. These neurons signal things like "I'm well fed" or "I'm starving" or, more generally, "internal conditions are good" or "they're not good." These signals affect how much attention or weight should be given to various external stimuli, tuning behavior accordingly; for example, a hungry animal will be more likely than a well-fed animal to cross a noxious chemical to get to food. These kinds of signals also inform decisions to keep on exploiting the current environment or to move on and explore somewhere else. In this way, the organism—*as a unified whole*—adapts its behavioral choices to align with a holistic view of its current situation and its goals.

But that is not all. This more complex architecture also provides the opportunity for individual organisms to learn and adapt over their lifetimes, using the past to predict the future and guide behavior. *C. elegans* are capable of many different types of learning, some that only apply over short timeframes and others that are very long-lasting. There are the simple forms—habituation and sensitization to single stimuli—seen in some of the other creatures we encountered so far. These allow the animals to change their responsiveness to something in the environment depending on recent exposure, whether its concentration is constant or changing, and so on. Useful but not super fancy.

Fancier is the ability to *associate* different things together—to learn that when you detect A, that means B may also be present or may shortly follow. Worms can be trained, for example, to prefer certain odors over others if they are paired with food or to avoid some odors, even ones they are innately attracted to, if they are paired with a noxious stimulus. This kind of learning and the resulting memory (not a conscious psychological memory of the type we experience, but a long-lasting cellular change reflecting the experience) can be mediated by changes in many different cellular parameters that effectively reconfigure the dynamics of the neural circuit. These changes include increasing the amount or sensitivity of receptor proteins for specific signals, up- or downregulating the electrical excitability of sensory neurons or interneurons, and altering the strength of connections between specific neurons so that a downstream neuron upregulates the weights of some signals relative to others. As the designers of artificial neural networks know, a hierarchical architecture, in which the weights of individual connections can be altered, is perfectly suited to learning all kinds of associations. We will see later that the deeper the network—the more levels it has—the more abstract the concepts and categories it can learn about.

In worms, all these mechanisms allow the *individual* organism to reconfigure its pre-wired circuitry to reflect knowledge about the world it encounters over its own lifetime and then to apply this knowledge in its own decision making. Signals about the animal's internal state are crucial in guiding these modifications. For example, worms raised at a specific temperature, where they were consistently well fed, subsequently demonstrate a long-lasting preference to seek out and remain at that specific temperature in a test environment. Worms raised at the same temperature, but under starvation conditions, show no such preference. Signals about their internal state, previously used to guide behavior moment by moment, later also act as *learning signals* that guide *future behavior*. They allow the individual to learn that certain environmental conditions are good or bad relative to its goal of keeping its internal state in a healthy operating range.

As we proceed in our evolutionary journey, we will see that these same signals are used in an even more sophisticated way: to evaluate the outcomes of actions. When an animal performs a certain action in a given situation, these signals about the resulting change in internal state will feed back to the decision-making systems and either reinforce or inhibit the choice of that action when the animal finds itself in a similar situation again. This goes beyond associative learning ("this external signal is predictive of conditions that are good for me") to reinforcement learning, giving feedback about the agent's own behavior ("in this context this action had a good outcome—if I encounter that situation again, I should do that again").

*C. elegans*, which have the ability to develop knowledge about their environment based on their own experience, highlight an increase in the degree of agency over what we encountered so far. We saw that simple unicellular creatures are biochemically configured to behave in ways that favor their own persistence and to respond in adaptive ways to their environment. Simple multicellular creatures show the same kind of adaptations, preconfigured and coordinated at scale by neural circuitry. These organisms have a repertoire of possible actions and choose between them *for reasons*.

But it could be argued that they are *natural selection's reasons*, not those of the individual organisms themselves. They come pre-wired, thanks to the life-or-death feedback of natural selection across preceding generations. What we see in *C. elegans* is a major step beyond that. Individual worms can learn from their own experience and develop *their own reasons* for choosing one action over another in any given situation. An individual worm is no longer just an instance of an evolutionary lineage—a preprogrammed drone rolling off the factory conveyor belt. It goes out into the world and develops *its own agency*, through the history of its own actions and its own experiences.

## Summary

Life got big and it got complicated. The acquisition of mitochondria gave new eukaryotic life-forms the luxury to increase in complexity. This opened the door to multicellularity, with a division of labor among diversified cell types. Coordinating this new type of body required muscles and neurons to control them. Neurons proved to be the perfect vehicles to link sensory information to action, and the evolution of a hierarchical architecture enabled greater integration of internal and external signals to guide behavior; in turn, this enabled the colonization of more complex and changeable environments. Finally, the emergence of associative learning and long-lasting memory let individuals transcend their pre-wired instincts and be able to make decisions based on their own reasons. The next step was to give them more *to reason about*.

# 5

# The Perceiving Self

The creatures we have met so far are capable of some impressive behavioral control and even what we can begin to call cognition. They can integrate multiple signals from the environment at once, adjust their responses based on their own internal states and changing goals, and even learn from experience to guide future actions. But they don't really have much cognitive *depth*. If they're thinking, they're not thinking *about* much. They cannot do that: their powers of perception are simply too crude. They perceive the world through smell and touch, which means the information that their nervous systems carry is simple, local, and immediate. The invention of hearing and vision, especially as animals ventured onto land, greatly expanded perceptual horizons, generated additional opportunities to inform action, and made more complex cognition a worthwhile investment.

Smell and touch provide crucial information to a behaving organism, but that is necessarily information about what they are literally in contact with. This property is obvious for touch, but even for smell the receptor proteins are activated by directly binding the molecule they are sensing. The receptors thus signal the presence of odor molecules only as the organism intercepts them. To guide behavior, the organism needs to make some inferences about how such molecules are distributed in the environment. To do so, signals can be compared between receptors on one end of the organism and the other, or they can be integrated over time as the organism moves, driving motion up or down a concentration gradient of the molecule being sensed.

Using these approaches, organisms build up a kind of internal model or map of what is out in the world. Or, to be more precise, their internal organization comes to physically reflect and is correlated with what is out in the world. If someone were able to see what was going on inside the organism (a scientist, for example), she might be able to infer based on the observed internal pattern something or even a great deal about environmental conditions. But the organism itself doesn't actually look at an internal map and then decide what to do. Instead, the responses to various signals are worked out pragmatically, through the operations of the relevant biochemical or neural pathways. The organization of those processes very literally embodies information about the outside world and involves the first glimmers of what is called *representation*.

To understand representation, we can think of the difference between a thermostat and thermometer. The physical state of a thermostat reflects the temperature of the environment, but it doesn't *represent* it and communicate it to something else. Instead, the temperature has a direct effect on the components of the thermostat itself. It may involve, for example, the uneven expansion of two sheets of different types of metal fused to each other, which results in bending of the composite structure and the making or breaking of an electrical contact, which then controls whatever the thermostat is in charge of (like turning the heating or the air conditioning on or off).

A thermometer does something very different. It too has physical components—a column of mercury, for example—that are physically correlated with and thus reflect the temperature of the environment. However, its function is not to directly act on that information but rather to represent it—to make the information about the temperature present and available to someone or something else. That information may be read off by a person or converted to a digital signal that is transmitted to other parts of a machine or technical system, where it may be combined with information about other things. The point of gathering information about temperature is still to guide action, but this is no longer achieved through direct coupling. This kind of system enables greater flexibility of response: you may want to also know the humidity, for example, before you decide whether to turn on the air conditioner.

In simple organisms that get around by smell and touch, this kind of internal representation happens, but the cognitive operations carried out on the represented information are fairly simple. For example, in *C. elegans,* the signals from various sensory neurons represent the presence of noxious chemicals or excessive heat or aversive touch; that is, they *report actionable information* about these stimuli to the rest of the nervous system. The activity of integrating hub neurons represents the simple sum of these aversive stimuli while discarding any information on what they are. The levels of aversive stimuli can then be compared with signals of attractive stimuli in yet other neurons to build up an even more complete picture of both good and bad things near the animal. This picture is ultimately transformed into a decision mediated by activation of command neurons that directly drive an appropriate action.

*C. elegans* do not need many levels of internal processing because they don't have that much to think about. They do build up a kind of internal map of what is outside in the environment, where it is relative to their own bodies, and how it is changing over time. But this map is highly selective, including only those things that the organism directly touches and those molecules that can be directly bound by its repertoire of receptor proteins. And it extends only a short distance in both space and time. It relies on active exploration and is thus limited by the speed at which the organism can travel and the transience of the detected chemicals themselves. From a sensory and cognitive point of view, these organisms inhabit a world of the here and now.

## Broadening Horizons

The evolution of hearing and vision opened up new vistas and, crucially for our story, gave organisms that developed these faculties much more to think about. Hearing is really a modified kind of touch—or *mechanosensation.* But instead of responding to direct contact by objects, it signals disturbances in the surrounding medium—air or water. This is a good trick, because the kinds of objects that an organism is likely interested in—potential prey or predators, for example—are often themselves moving around and thus generating vibrations that can be detected when the object is still some distance away.

Objects in the environment also create disturbances in the ambient field of electromagnetic radiation, which can similarly be detected and used by an organism to inform its behavior. Visual systems have evolved to detect and parse these disturbances: gradations and discontinuities in the different frequencies and intensities of light (in the visible range of electromagnetic radiation) that are reflected by different kinds of objects, located in different positions in the visual field.

Both hearing and vision have evolved to allow organisms to *infer the sources* of the detected stimuli and where they are in the world, thereby building up a map of objects around them. From the outset, these aspects (what, where) were linked with information about whether the inferred stimulus was good or bad (to be approached or avoided), what should be done about it, or what could be done with it. Like smell and touch, these senses started with direct couplings to action. But, as we will see, they offered much greater evolutionary scope for elaboration and increasing depth of internal processing. Because visual systems have been studied in much greater detail, we will concentrate on them.

## Sniffing out the Light

Many creatures, including some kinds of bacteria and many single-celled animals, can detect and respond to light. These systems all rely on specialized *photoreceptor* proteins, called *opsins*, that can absorb photons of light; this absorption changes the conformation of the protein in some way that activates a cellular response, either by triggering an internal chemical signaling cascade or directly opening an ion channel in the membrane. These opsin molecules work in a surprisingly similar way to odorant receptors. They do not react to light directly. Instead, they bind a small molecule—called *retinal*—whose chemical conformation is altered when it absorbs the energy of a photon. When it is photoconverted in this way, retinal is released by the opsin protein (at least that is how the opsins in our own eyes work), resulting in a knock-on change in the conformation of the opsin and a resultant chemical cascade. The photoreceptors are thus really chemoreceptors, like odorant receptors, but for a molecule that is itself sensitive to light. In a sense, then, these proteins indirectly "smell" the presence of light.

In the simplest examples, these photosensitive systems do not produce what we would call vision, in the sense of forming some kind of image of the distribution of objects in the outside world. Instead, the earliest evolving systems—still extant in many organisms—are only used to monitor and respond to the absolute levels of ambient light. This function can be useful, for example, for keeping an organism's internal circadian rhythms entrained to the environment (something we still do ourselves).

Some kinds of marine plankton—the term for all types of small creatures that drift with ocean currents—track ambient light levels to regulate their depth in the ocean, depending on how intense the sunlight is. Many such creatures are capable of converting the sun's energy into cellular energy through photosynthesis and so must move to a depth with sufficient levels of sunlight to live, which will vary with atmospheric and marine conditions, time of year, time of day, and so on. But they also must avoid the highest intensities of solar radiation, which can have harmful effects. Because these creatures are food for other organisms, many of their predators have similarly evolved mechanisms to track ambient light levels, which very handily predict where their lunch can be found.

An even better trick is to create an organ that allows it to sense the *direction* from which light is coming. The simplest versions are found in various single-celled creatures in the form of an *eyespot*: a cluster of photoreceptor molecules on the cell surface, which is shaded on one side by a patch of light-absorbing pigment. The pigment is crucial in this design because, while the photoreceptors are being activated, there must be light coming from the nonshaded direction. As the organism moves around, the way the light intensity varies with respect to the position of the pigment can be used to identify the direction it is coming from. By linking the resulting signals to motor systems (which can also integrate information about internal states), the organism can regulate its movement toward or away from light (see Figure 5.1).

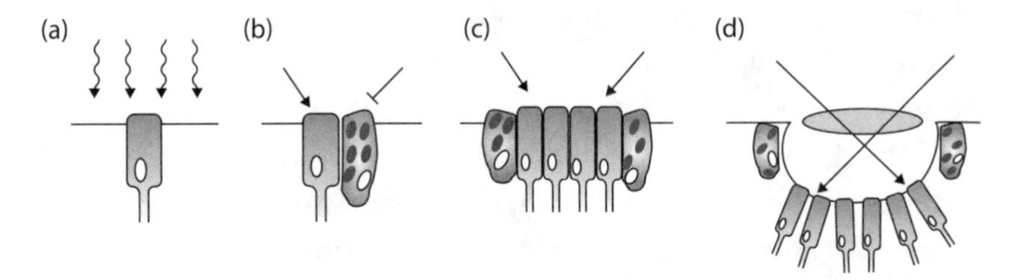


FIGURE 5.1. Evolution of vision. (**a**) Simple photoreceptor cells enable organisms to respond to ambient light levels. (**b**) Pigment-bearing cells can shield light, enabling directional photoreception. (**c**) Arrays of photoreceptor cells provide low-resolution vision, with different photoreceptors detecting light coming from different parts of the visual world. (**d**) The evolution of a focusing lens and the arrangement of photoreceptors into a retinal array enabled high-resolution image formation and processing through subsequent layers of neural circuitry. Each step along this evolutionary trajectory went hand in hand with the ability and need to carry out more complex behavioral tasks.

So far, these responses are similar to the kinds of chemotaxis (moving toward or away from various chemicals in the environment) that we saw in bacteria and Paramecia. But light signals can also be used to trigger much faster responses. These include rapid, automatic escape reflexes as well as much more sophisticated mechanisms of image processing that have evolved in many kinds of animals, including fish, insects, and mammals. To track these developments, we must return to our evolutionary tree.

## New Lifestyles

As we move past the branch of the tree that gave rise to Cnidarians, such as Hydras and jellyfish, we come to a point where the common ancestor of much more complex creatures—everything from insects to octopuses to you and me—must have existed. We don't really know what this supposed creature looked like, but from what those diverse descendants have in common, we can make some educated guesses.

It probably had a fairly simple body shape, like a flatworm, with the important innovation of bilateral symmetry. That is, unlike sponges and jellyfish and the other earlier-evolving creatures we discussed so far, which are radially symmetric, this flatworm had a left and a right side, as well as a head and a tail, and a back and a belly. Indeed, the group of creatures from this point of the tree on are collectively called Bilateria. They comprise multiple branches, including nematodes (like *C. elegans*), arthropods (including insects and crustaceans), annelid worms (like earthworms), mollusks (including octopuses, squid, snails, slugs, and all kinds of shellfish), and chordates (which include vertebrates—fish, birds, amphibians, reptiles, and mammals).

Although these organisms have flamboyantly diverse morphologies, their basic, common body plan is a cylinder surrounding a *through-gut*, with a mouth at one end and an anus at the other. Muscles are distributed around the body wall, down the length of the cylinder, and the nervous system runs in a tube centrally, with a condensation in the head, which we call the "brain." From that basic body design have emerged all kinds of specializations: different ways of getting around, sensing the world, and interacting with it.

Of particular relevance for this story is the geological period known as the Cambrian (which lasted from 540 to 485 million years ago). It witnessed an explosion in new forms of marine life, with all kinds of fantastical creatures suddenly evident in the fossil record (if emergence over 55 million years can be called sudden). Most of these creatures are now extinct, but this period did produce the ancestors of all the groups of the more complex Bilateria. Why there was such an apparent explosion in animal forms at this time is much debated; part of the answer may have to do with the types of sedimentary rocks being laid down at this geological time, which are ideal for fossilization. But recall the general principle that innovation drives more innovation.

Evolution is an inherently creative and unpredictable process, but it is more than a series of "frozen accidents." Natural selection combs through new variants for ones that survive and reproduce better than their peers. Sometimes that is achieved by doing the same things a little bit better. At other times it is achieved by doing something new. And every time that happens, every time some members of a species take a step in a new direction, it opens up new possibilities and creates new niches that can be exploited by other creatures. Stuart Kaufmann, a theorist of complex systems, refers to this as exploring the "adjacent possible" and compares this dynamic to that seen in economies in response to technological innovations. Each new technology—steam engines, computers, smartphones—creates entirely new, often unforeseen opportunities that are rapidly taken advantage of by entrepreneurial individuals. Nature similarly exploits such prospects for making a living.

Throughout this massive diversification of forms and lifestyles, the means of perception coevolved with the means of action, according to the demands and opportunities of the niches that these new creatures adapted to exploit. These new opportunities likely started on the sea floor, as creatures went from swimming to crawling and from scavenging to actively hunting. The bilateral body plan is well suited to this kind of lifestyle, and our hypothesized flatworm-like ancestor (called Urbilateria) was likely capable of crawling around and looking for tasty titbits, directed by information from its senses that may have included primitive eyes.

Explaining how eyes evolved, of the type we see in modern animals with their array of specialized parts and exquisite apparent design, was seen as a problem for the theory of natural selection, one that even Charles Darwin admitted. It wasn't obvious how a gradual accumulation of these parts could have happened nor what use any of them would have been until the whole machine was assembled. But as we have seen, even simple clusters of photoreceptor molecules are extremely useful and can be elaborated on in a gradual fashion as different needs dictate.

Another concern was the supposed implausibility of eyes evolving multiple times, independently, across the diverse lineages of animals that possess them. There is a striking multiplicity of detailed eye designs across insects, crustaceans, mollusks, and vertebrates, and many other animals lack eyes altogether. But rather than this variety signaling independent origins, genetic data suggest that some primitive kind of eye was present in Urbilateria and evolved in different ways across different lineages, including many instances of loss when vision became dispensable. The same genes are used to build eyes across all these groups of animals, indicating the presence of this developmental system in the common ancestor. Clearly, eyes were a good idea, one that enabled all sorts of lifestyle innovations.

## Something Wicked This Way Comes

One function eyes serve is as an early warning system. While you are wandering about looking for food, other things may be looking for you. A passing shadow may signal the presence of such a predator, and our early ancestors evolved systems to detect and react to such stimuli. A single eye can create a warning signal by the direct coupling of nerves coming from it to neurons controlling a simple, explosive escape response. Some insects use specialized eyelike organs, called *ocelli*, which sit on the top of their heads, for this purpose, initiating a rapid jump reflex in response to a shadow passing overhead (which is partly why it's so hard to swat a fly).

But two eyes provide an advantage in that they allow *directional* movement, away from the threat. In the simplest versions of such escape circuits, which are still present in many fish, nerves from each eye cross the midline of the nervous system and innervate a bilaterally paired behavioral control center in the midbrain, called the *tectum*, on the other side. Nerves from the tectum on that side then project down the spinal cord to initiate muscle contractions on the same side of the body. A shadow approaching from the left causes the fish to contract the muscles on the right side of its body and thus turn away from the threat.

The circuits subserving these basic kinds of actions have been well studied in the lamprey, a simple jawless fish, by Sten Grillner and colleagues. Directly electrically stimulating different parts of the lamprey tectum can not only induce turning, as described, but also rapid swimming or wriggling. In addition, stimulation of one subregion activates approach, rather than avoidance. This region is innervated by neurons from the eyes that are focused on the region directly in front of the animal, consistent with their role in predation. The lamprey tectum thus contains a kind of action map that is analogous to the command neurons we saw in *C. elegans*. And as in *C. elegans*, these different sets of neurons are coupled directly to different sensory inputs.

We will look more at how actions are selected in the next chapter. For now, the important point is that the meaning of these signals is grounded in action and the verdict of natural selection. Evolution has burdened living things with a glorious purpose: to persist, either in their own right or through reproduction. In the service of this overarching imperative, animals evolved subgoals: find food, avoid threats, and mate. Different stimuli in the world have value and relevance *in relation to these goals.*

For perception to guide behavior, it has to provide the answer to two questions: What is out there, and what should the organism do about it? Surprisingly, evolution seems to have answered the second question first. A shadow need not be separately apprehended as such. It doesn't have to initiate a signal that means "here is a shadow." What it does have to do is initiate a signal that, for the organism, means "danger!" and prompts a response to correct this undesirable scenario. If the animal escapes, then the verdict from natural selection is a thumbs-up, and the reward is survival for another day and ultimately a greater likelihood of successfully reproducing. The overall effect is that escaping from shadows, as a *control policy,* will become selected for: the genetic variations that cause some individuals to tend to take this action in response to real threats in the environment will outcompete others and may even become fixed in the population.

These meanings are pragmatic and not yet internalized. But of all the senses, vision offered the best opportunity for moving beyond these kinds of pragmatic couplings to true internal representations: signals that carry decoupled information on what is out in the world.

## What's Out There?

Visual abilities became more and more useful as life moved on land. The first creatures to make that move were probably arthropods, the ancestors of insects. But they were followed very quickly by some ancient bony fish, which likely already possessed simple forms of both lungs and limbs. These creatures evolved into amphibians capable of breathing both air and water. They had the land to themselves for a while but couldn't venture too far from water because they had to lay their eggs there. Eventually, new forms emerged that would become reptiles, with the key innovation of an egg with a protective shell that prevented the embryo from drying out. This new type of egg let these creatures move farther from the water to explore new ecological niches, prompting further diversification.

These early reptiles gave rise to several branches, including one that leads to current reptiles like alligators and snakes, one that evolved into dinosaurs and then birds, and one that led to the earliest mammals. One of the major distinguishing characteristics of mammals is that they are warm-blooded: they actively expend energy to keep their bodies at or very close to a specific temperature. This gives them much more freedom from the vagaries of the external environment, allowing them to move around at night, for example. But this feature comes at a steep price: mammals use a huge amount of energy just keeping their body temperature constant. This meant even greater pressure to be efficient predators, placing an extra premium on sensory abilities, especially because the major prey were probably fast and tiny insects.

The first mammals were themselves quite small (shrew-sized) and nocturnal and may have hunted mainly by smell. But when the dinosaurs became extinct, they left some job openings that mammals swiftly filled. Many became diurnal (moving about during the day), got bigger, and started hunting bigger things, including other mammals. Vision became a good investment again.

The visual systems that most animals possess are designed to enable much more than simple responses to variation in light levels. Differences in intensity and wavelengths of light bouncing off different objects in the environment can be used to infer the presence of objects out there in the world, what they are, where they are relative to each other and to the organism, and how they are moving—all invaluable information for a creature navigating through life.

Detecting these patterns of light is the first task that the visual system must do, but just capturing an image, like a camera, is not the purpose of vision. To be useful in guiding action, seeing requires parsing, segmenting, color coding, labeling, recognizing, and tracking objects through space and over time. This is vastly more complex than what goes on when you snap a picture with your smartphone; it is more similar to all the work that a self-driving car has to do to get around safely. These processes begin with circuits in the retina of the eye itself and continue through a complex hierarchy of brain regions devoted to vision.

The first step is to arrange a bunch of photoreceptive cells in an orderly array, usually closely packed in a two-dimensional sheet. The most primitive eyes probably were just sunken cups with such arrays on their inside surface, surrounded by light-absorbing pigment that served the same shading function as in eyespots in single-celled creatures. Different cells in this eye cup would then be affected by rays of light coming from different parts of the environment. The subsequent evolution of a lens, which could both capture more light and focus it onto the photoreceptive cells, made the individual photoreceptors even more locally selective to light from different parts of the visual field (see Figure 5.1).

Having a two-dimensional array of photoreceptor cells, each responsive to light from neighboring spots in the visual world, immediately suggests a way to parse the image—by comparing the activity of neighboring cells. This function can be accomplished with a second layer of cells, each of which takes inputs from a number of neighboring photoreceptors. If you want to distinguish different objects, the best way to do so is through contrast enhancement, parsing the image into regions of constant lighting versus lines and edges where the lighting suddenly changes. That is exactly what the multiple layers of neurons in the retina do.

To enable contrast enhancement, the activity of each photoreceptor needs to be converted into both excitatory and inhibitory signals. This conversion is carried out by a layer of "bipolar cells," which come in two types: "ON" and "OFF." The activity of the bipolar cells is monitored by another layer of neurons, called *retinal ganglion cells* (RGCs). Each RGC receives inputs from multiple ON bipolar cells that collect information from a central set of overlying photoreceptors. The RGC is thus activated when these photoreceptors are activated. But the RGC also receives inhibitory inputs from multiple OFF bipolar cells, which are responsive to neighboring photoreceptors and convert these signals into an inhibitory influence on the RGC. Thus, an RGC of this type is activated most strongly when there is light coming from a particular spot in the visual field while *less light* comes from its immediate surround. (There are also RGCs wired in the opposite way: they are more sensitive to light in the surround than in the center) (see Figure 5.2a).

The RGCs thus act as contrast detectors for light intensity. A broad region of constant lighting activates an RGC of this type less than an edge where the properties of the light change. Other RGCs do the same for detecting contrasts in the *wavelength* of light coming from different regions. In our retinas, there are two types of photoreceptors: *rods,* which respond very sensitively to light across a wide range of wavelengths, and *cones,* which are tuned to selectively respond to light of specific wavelengths (because they express different opsin proteins). The cones come in three varieties, each sensitive to a different peak wavelength: red, green, and blue light. A similar architecture of bipolar cells and RGCs carries out a comparison of the wavelength of light impinging on neighboring cones in the retina, comparing the intensity in the red versus the green channel, or the intensity in the blue channel versus the sum of the red and green channels, to produce a blue-yellow contrast.

This process provides an additional means of enhancing contrast between nearby objects that often reflect light of different wavelengths. Indeed, there are no discrete colors in the world. They are entirely a creation of our perceptual system, literally designed to color code the variety of objects we encounter in the world (enhancing the *perceptual contrast* between ripe fruits and green leaves, for example). Different species do this in different ways, detecting a different range of wavelengths (some into what we call the infrared or ultraviolet ranges) and comparing channels selective for different wavelengths. People who suffer from color blindness (often due to genetic mutations affecting the opsin proteins that are sensitive to different wavelengths) have much more limited ability to make these discriminations.

The retina contains dozens of different types of RGCs, all specialized for parsing different kinds of visual information—high or low resolution, colors, movement, flicker, and so on. In the human retina there are about 1.2 million RGCs but 125 million rods and cones. Each RGC thus integrates information from around 100 photoreceptors. The

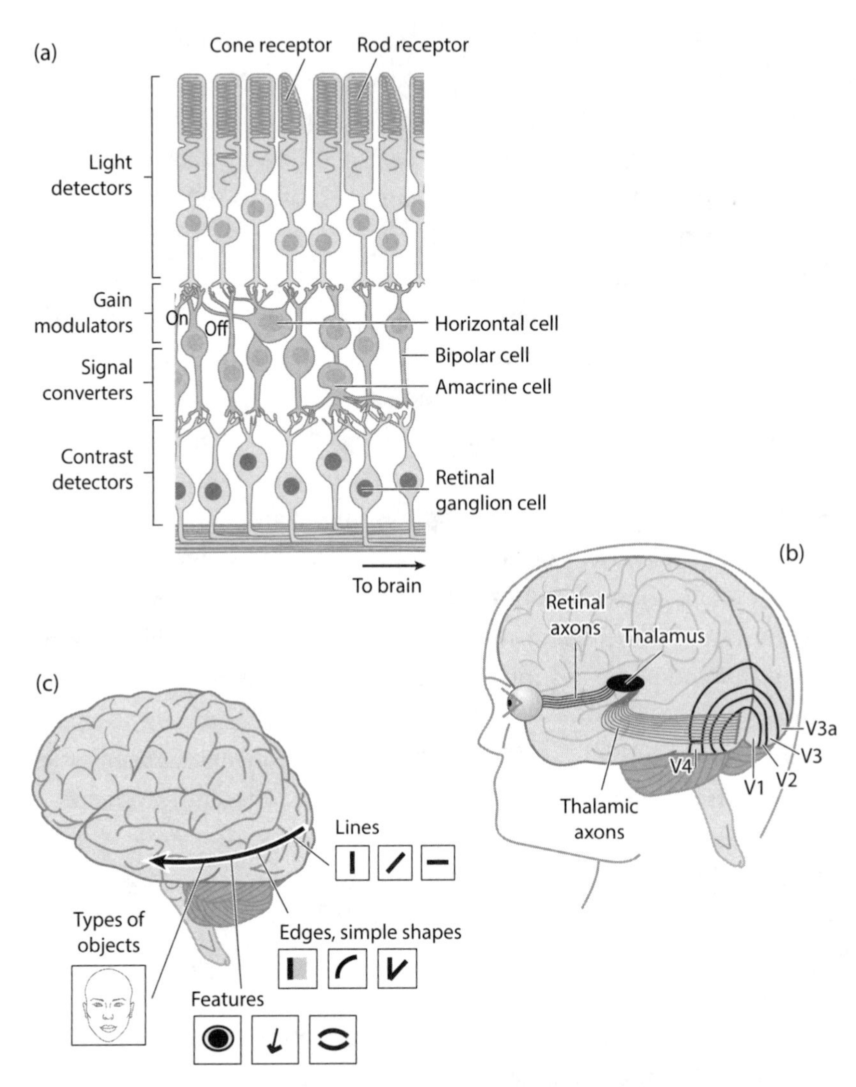


FIGURE 5.2. Visual processing. **(a)** Signals are processed through the layers of the retina to extract the most meaningful elements of the visual scene. **(b)** Retinal axons project to the thalamus, and thalamic axons project in turn to the primary visual cortex (V1), and so on through 30 different visual areas (V2, V3, V3a, V4, and many others). **(c)** Higher-order visual properties are extracted through this hierarchy, from lines to simple shapes to features of objects to specific types of objects, such as faces.

RGCs are also the output neurons of the retina: they send a long projection along the optic nerve to convey the resulting signals into the brain. It is important to note just how much information processing has already happened by this stage and what it entails: salient features are extracted, while irrelevant information is discarded—averaged, filtered out, effectively ignored. The retina thus extracts ecologically meaningful information from the barrage of light impinging on the photoreceptors and sends only these relevant signals to the brain.

## Decoupling Perception from Action

In fish, birds, amphibians, and reptiles, the retina projects mainly to the tectum—the midbrain area we saw in the lamprey that collects incoming sensory signals to inform the choice among possible actions. The tectum also receives other sensory information—tactile and auditory, for example—which is integrated with visual signals to generate a holistic picture of the current environment relative to the animal. In mammals, the tectum is known as the superior colliculus, and it still performs similar functions, guiding actions like automatic gaze shifts toward movement in the visual periphery. Its primacy in behavioral control has been superseded, however, by the evolution of the neocortex.

The neocortex is the region at the front and top of the forebrain, which expanded greatly in mammals, especially along the lineage leading to humans, in whom it makes up three-quarters of the mass of the brain (see Figure 5.2, lower panels). It is so large that it has to be folded in on itself to fit inside the skull, like crumpling a big sheet of paper to try and fit it inside a tennis ball. In evolutionary terms, the neocortex is part of an outgrowth of a more ancient structure called the *hypothalamus* (see Figure 5.3).

The hypothalamus is a vital control center in the forebrain; indeed, in the earliest chordates, it probably comprised almost the entirety of the forebrain. Its ancient functions—still active in us—mainly involved regulating bodily physiology in response to changing environmental or internal conditions. This regulation is achieved by electrical neural signaling and by the release of various hormones affecting appetite, blood

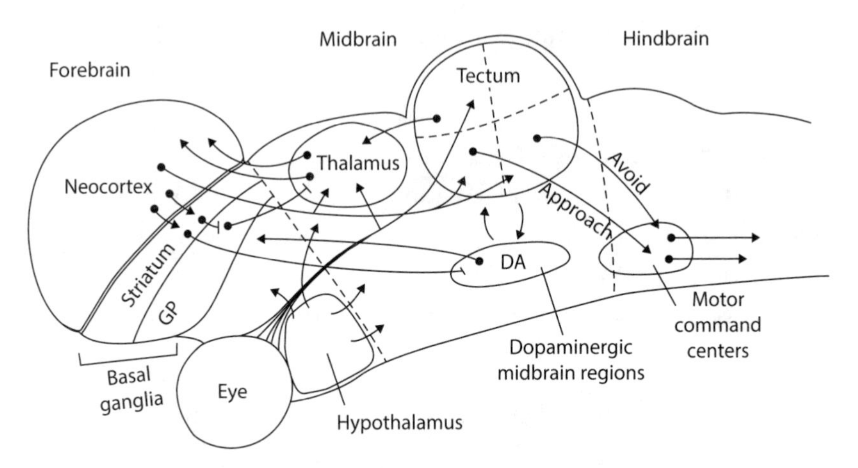


FIGURE 5.3. Evolution of neocortex. The major structures of the forebrain (including the neocortex and basal ganglia (striatum and pallidum)) evolved as outgrowths of the hypothalamic region. These regions integrated with existing structures in midbrain and hindbrain involved in behavioral control, including the thalamus, tectum, dopaminergic midbrain regions, and motor command centers, to provide an additional level of cognitive control of behavior.

pressure, reproductive physiology, and other *autonomic* functions: all those aspects of physiology that are regulated by the brain without conscious awareness. In primitive chordates, the hypothalamus also likely controlled the crucial decision whether to exploit currently available resources or explore elsewhere.

This decision was informed by signals such as dopamine, which indicated the current internal nutrient state. But early in chordate evolution, olfaction began to be used as an additional source of information about the availability of nutrients *outside* the organism. An expanded hypothalamus evolved to process this information and integrate it into decision-making circuits. This was the early *telencephalon* (meaning the end of the brain), which would ultimately expand in mammals to become the neocortex and other structures we explore more in the next chapter. Its basic functions remain the same as that of the hypothalamus: to integrate external and internal signals about things in the world and the state of the organism and to use them to inform decision making. But the structures of the telencephalon, especially the neocortex, allow these functions to be carried out with much more subtlety and sophistication than the hypothalamus.

The architecture of the neocortex is especially well suited to this role, because it has a modular, repeating columnar structure that can readily be scaled up. The neurons in the neocortex are organized into six layers, spanned by these columns, with circuitry specialized for various kinds of computations (see Figure 5.4). In simplified terms, within a given column of cells, layer 4 receives inputs from other parts of the brain, including ones carrying sensory information from a region known as the thalamus—more on this later. Neurons in layers 2 and 3 collect this information and integrate it with inputs from neighboring columns or other cortical regions. Neurons in layer 5 take the resulting signals, along with contextual information from other cortical areas, and send outputs to other parts of the cortex, including in the other hemisphere, and to other parts of the brain, notably the tectum and spinal cord. Neurons in layer 6 send crucial feedback signals primarily back to the thalamus. Interspersed with all these excitatory neurons are diverse classes of inhibitory interneurons, which help sculpt the dynamics of information flow through these circuits.

Each column is thus a functional unit designed to take in a variety of inputs, operate on them in some way, communicate with neighboring columns to adjust its own activity, and send outputs to other parts of the cortex and the brain. Hundreds of millions of such columns are vertically arrayed parallel to each other to make up the entire cortex, which can be thought of as a two-dimensional, multilayered sheet. Or perhaps it is more like a patchwork quilt, because the cortex is also divided in the horizontal domain into many discrete areas with specialized functions.

In mammals, many of these areas are devoted to vision. Retinal ganglion cells project their axons along the optic nerve into the center of the brain, branching to innervate the tectum and also the thalamus, the *inner chamber* of the brain. The thalamus is generally thought of as a simple relay station for visual, auditory, and tactile information coming from the sense organs. But it also performs significant processing on those inputs, pooling inputs from neighboring sensory cells to remove

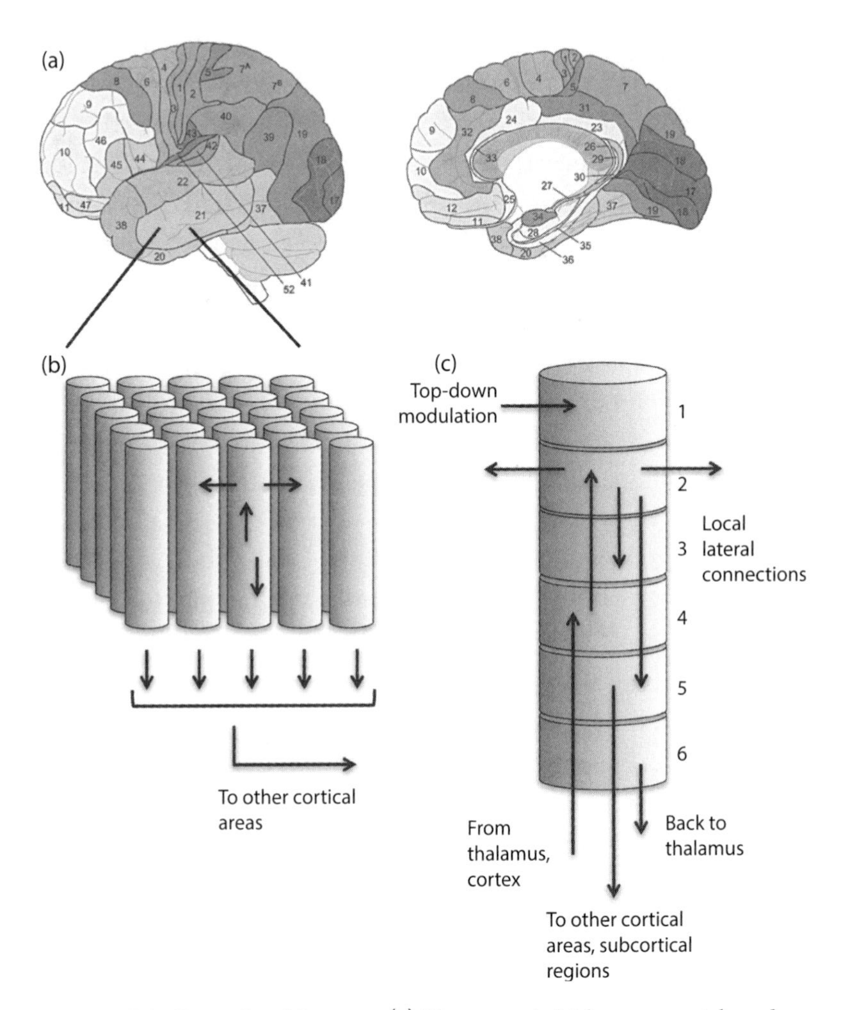


FIGURE 5.4. Cortical architecture. **(a)** The cortex is laid out as a patchwork of areas that can be distinguished based on *cytoarchitecture* (the microscopic organization of different types of neurons within the area) and connectivity (which regions they receive inputs from and which ones they project to). **(b)** Each area is composed of densely packed cortical columns, each containing thousands of neurons arranged in stereotyped layers. **(c)** The neurons in different layers play distinct functional roles, receiving different kinds of inputs, forming connections between layers in the same column or with neurons in neighboring columns, or projecting further to other areas of the cortex or the rest of the brain.

redundancy and compress the signal before sending it on to the cortex. Thalamic neurons project to the cortex in a highly specific way, with different subdivisions of the thalamus projecting to different areas of the cortex.

In this way, information from the retina is conveyed, via the visual subdivision of the thalamus, to an area called the primary visual cortex (or V1) at the very back of the brain (see Figure 5.2, lower panels). This is done in a highly structured fashion that crucially maintains nearest-neighbor relationships; that is, neighboring cells in the retina, which respond to light from adjacent points in the visual field, project to neighboring cells in the thalamus, which project to neighboring cells in V1. The result is a map of the visual world, as it impinges on the retina, across the two-dimensional extent of V1. In fact, V1 contains multiple such maps, because circuits carrying different kinds of information from the retina (related to form, motion, and color, for example) remain segregated in the thalamus but then become intercalated within V1.

The result is that individual neurons in V1 receive inputs from multiple thalamic neurons that convey information from adjacent points of the visual field. Using similar mechanisms as in the retina, these inputs can be summed or subtracted to infer higher-order relationships. For example, many cells in V1 are configured to be responsive only when multiple neighboring thalamic inputs aligned along a certain direction are active. This makes them excellent *line detectors*; in fact, many V1 cells selectively respond only to lines of specific orientation. Such lines are of interest because they often correspond to edges of objects, and objects are what the organism cares about.

At the same time, information on color and motion is being similarly processed to extract higher-order features. In addition, inputs from the two eyes become integrated within V1, with the slightly offset visual perspectives providing essential depth information. All this information is extremely useful in segregating figures from the visual background. However, the portion of the visual field that each V1 neuron sums over is still pretty small, and much more processing is needed to usefully segment the entire image detected by the retina. This is where the scalable design of the cortex really paid off.

It turns out to be very easy to make the cortex bigger. The neurons in the cortex are generated from a population of stem cells that divide over and over again, producing neurons each time. Some simple genetic changes can alter the way they divide so that more neurons are generated. These give rise to new columns of cells, expanding the overall size of the cortex. For a variety of reasons—some to do with wiring efficiency—the separate functional patches of the cortical quilt can only get so large before they split into two. As the overall size of the cortex increases, new functional areas thus tend to arise.

In early mammals, most of the cortex was probably taken up by a primary visual area, V1, and similar areas for audition, A1, and somatosensation or touch, S1. As the cortex expanded, new areas were added in between these primary sensory areas that were then devoted to additional processing of the various senses, according to the ecological needs of each species. For animals relying on vision, this eventually led to a huge increase in the number of visual cortical areas and an entire hierarchy configured for the extraction of the most relevant features.

This function is achieved by doing more of the same kinds of computations over additional areas of the cortex, allowing the extraction of features over larger and larger parts of the visual world. Neurons from V1 project to a neighboring cortical area, called V2, which projects to V3 and V4, and so on. At each stage, higher-order features are extracted: neurons become responsive to shapes and then to larger objects and even become specialized for types of objects. In humans, those larger objects include the types of most importance to our survival: other people's faces (see Figure 5.2, bottom panels). In parallel to this stream, which figures out what is in the world and where, another stream analyzes movement and figures out what all those objects are doing.

All this information is eventually integrated with our memory systems (more on them later) so that we do not only segment the image into visual objects but also recognize what they are. We thus label the objects that we see, insofar as we can, and draw on our knowledge of their properties to help determine what actions we should take with respect to them.

It's worth pausing here to consider how far we have come from simple systems that directly, or at least proximally, couple particular sensory signals to particular actions (like coupling detection of a shadow to taking evasive action). All the visual information processing discussed here is decoupled from action. There are now levels and levels of internal processing in which information is being processed, parsed, and transformed from each cortical area to the next. The outcome of all this processing will, of course, eventually be used *to inform* action, but not until the organism knows what the incoming signals mean.

## Seeing Is Believing

And there the organism faces a serious challenge. All it receives is a pattern of light of different intensities and wavelengths impinging across the array of photoreceptors in the retina. The job of all those levels of visual processing is to infer what is causing that pattern. But there isn't one single answer to that question. Any given arrangement of things in the world causes a unique pattern. Yet the converse is not true: many different arrangements of things can cause the same pattern of light. An object could be small, or it could be far away. A visual line could be the continuous edge of one object or be two objects that happen to be aligned. Successive activations in nearby regions could arise from a single object moving or one object disappearing from view and another appearing.

The organism thus has to solve this "inverse problem" by making *inferences* about what is causing the detected pattern of light. All the various types of information—light intensity, wavelength, motion, depth, contours, shading—are drawn on to help progressively segment the visual image. Crucially, the organism is not passive in this process: it can move its eyes, its head, or its whole body in ways that change the visual image from different objects in different ways, giving additional clues to what and where they are. It also can use information from other senses to calibrate these inferences and correct them if necessary.

The result is thus not a processed image like a photograph but really a set of beliefs. What is *represented* by the patterns of neural activity at any level of the hierarchy—that is, what is reported or made available to another part of the system—is not a line or a shape or a face at a particular position in the visual field but *the belief that* there is a line or a shape or a face at that position. One reason to think of them as beliefs, as opposed to propagated and processed signals that correlate with things in the outside world, is that they can be wrong.

This is revealed strikingly in people who suffer from hallucinations. But it is also clearly shown by all manner of optical illusions, which have been studied by psychologists and cognitive scientists for well over a century to try and divine the operations that the visual system carries out. Some of these illusory effects—like the Kanizsa triangle, for example (Figure 5.5)—highlight just how strongly our visual systems try to segment the image into distinct objects. Where there is some ambiguity about what the object is—as with the Necker cube—our visual systems alternate between believing in the presence of one versus the other. We receive the same visual information, but we interpret it differently from moment to moment.

Optical illusions also reveal how strongly our perception is shaped by our expectations. Whether figures seem huge when transposed to the end of a corridor, or a square on a checkerboard looks brighter than it is because we know it is in the shade, we clearly calibrate our inferences based on context (depth and lighting, in these examples) and on what we know about the world from our past visual experience. The potency of top-down expectations is also very evident in the auditory domain, where ambiguous sounds can be interpreted (very clearly heard) as quite different words (like "brainstorm" versus "green needle") depending on which phrase the listener is thinking of. The famous *McGurk illusion* illustrates the same idea: the sound we hear from a video of a person speaking ("da" or "ga") is altered by our simultaneous perception of the shape their lips make (even though the audio remains the same).

These are not just exceptional cases or curiosities—they show us that perception is not a system of one-way traffic. Signals from our senses are not merely passed and processed from lower levels up to higher and higher levels. Where would this end? How would any of these levels

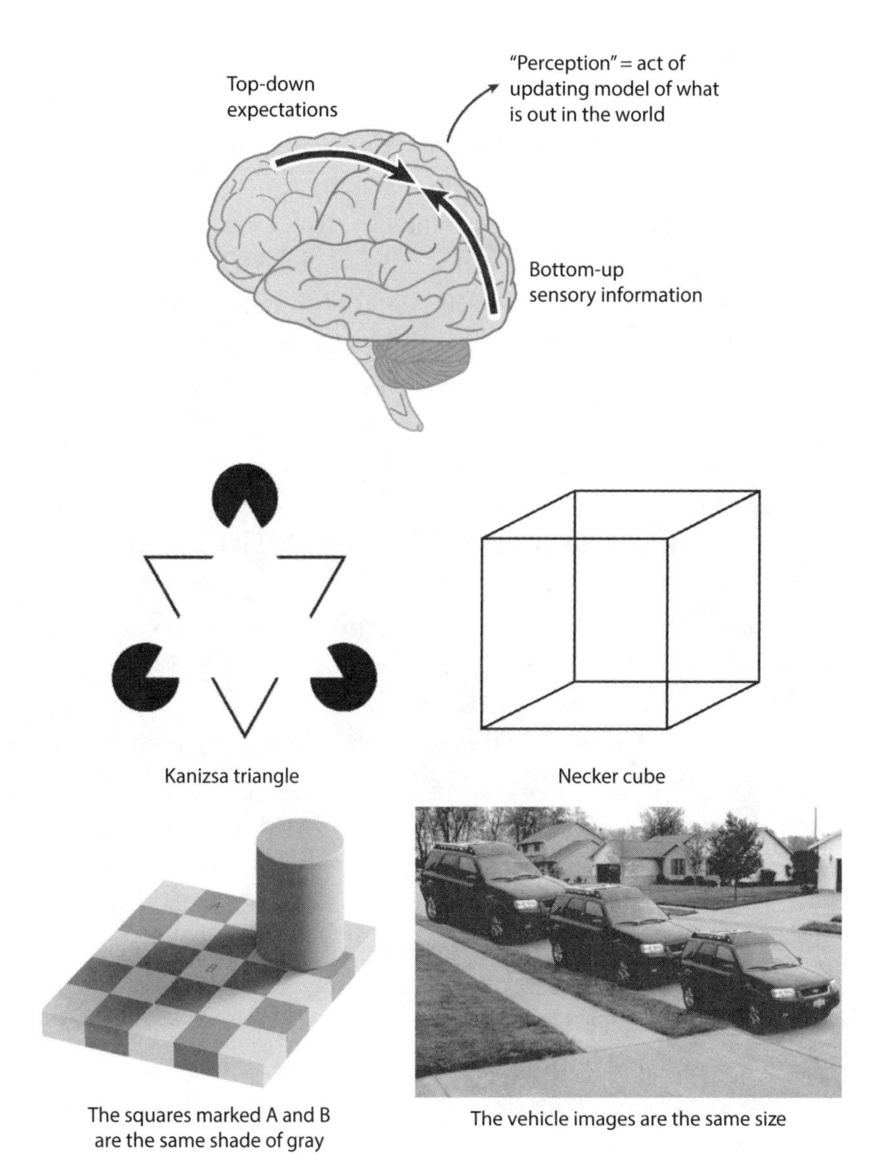


FIGURE 5.5. Perception as inference. Perception involves an updating of an internal model of the world, based on a comparison of incoming sensory information with top-down expectations. The Kanizsa triangle and Necker cube illustrate how our visual system tries to make sense of the world, inferring the existence of a white triangle that isn't there or insisting that the cube must be in one or the other orientation. Bottom panels show two illusions that illustrate the strength of top-down effects.

ever "see" anything? They don't, of course: it's the organism that sees things, and for us, what we see are the inferences we make. Those inferences are not solely directed by the signals from our senses but also are informed and constrained by our expectations.

The basic expectation—the ground on which we all stand—is that things in the world should continue to be pretty much the same from one moment to the next. That does not imply a static view because it includes the expectation that things that are moving should continue to move. Our current model of the world—what's in it and what's moving in what direction—thus furnishes an excellent predictor of what the next state of the world should be, especially when it is informed by stored knowledge of how things in the world typically behave (which is not necessarily consciously accessible but is embodied in the neural circuits).

A prominent hypothesis for how perception works involves a comparison of this model with incoming information. The idea is that higher levels in the hierarchy pass a prediction for the next state of things in the world (framed in the types of things that each level cares about) down the hierarchy through feedback projections; for example, from area V4 to V3, and from V3 to V2, and V2 to V1. At each stage, a neural operation is performed that compares the incoming sensory signals to the top-down predictions. Any discrepancies between them are detected and sent forward, up the hierarchy, so that the model at each level can be updated. In this framework, it is the updating of our internal models that constitutes perception.

When configured in this way, perceptual systems are not just processing information—they are *extracting meaning*. The patterns of neural activity across different areas in the visual hierarchy represent the system's best guesses of what is out in the world, focused on what is most relevant and important for the survival of the organism. Those guesses are not merely passively computed through successive levels of information processing. The organism is actively, subjectively interpreting this information, bringing its prior experience and expectations to bear. Indeed, it also brings the prior experience of all its ancestors to bear through the evolutionarily selected genetic preconfiguration of these systems in a way that is adapted to the regularities of the organism's environment.

## I Move; Therefore I Am

There is one final, important implication of this predictive, interactive, inferential view of perception. The goal of all this perceptual work is to build up an internal model of what is out in the world—especially to track what is moving, how fast, and in which directions—so the organism can anticipate things and not just react to them. But one of the things that may be moving is the organism itself. This creates an opportunity—to make inferences about the world by active exploration. But it also creates a problem: How can you tell whether changes in the visual image impinging on the retina are due to something out in the world moving or to you moving? If you've ever sat on a train at a station and had an adjacent train start to change position relative to you, you may have had the disorienting experience of momentarily not knowing whether you were the one moving or not. That kind of confusion would not do at all for an organism trying to track prey or avoid predators.

There are a number of solutions to this problem. The first is to incorporate expectations of how the world should change depending on how you are moving into the interpretation of the visual image. For example, if you are walking or running forward, everything in the visual scene should loom toward you, with closer things looming at a faster rate. Anything that is moving differently from that overall pattern is probably doing so under its own devices.

A more sophisticated approach is for the action system to send a message—a kind of internal copy of the action command—to the visual system every time an action is intended. The expected sensory consequences of this action are then actively subtracted from even the low-level sensory areas. This happens all the time and explains why when you move your head or shift your gaze you do not perceive the whole world as *moving*, even though the image on your retina shifts massively. Indeed, we make little subconscious eye movements—called *saccades*—several times a second, without perceiving movement in the world. This is possible because the visual system is forewarned about the impending movement and is factoring it into its predictions. By contrast, if you use your finger to (gently!) press your eyeball from the side to make it move a little, you will see the world shift in a disconcerting fashion. You might think that we could predict that too, but clearly our visual systems are not prepared to deal with the consequences of such an unusual action.

The upshot of having to distinguish self-caused from non-self-caused movements in this way is that the organism must include *itself* in its model of the world. The only way to productively make sense of things is for the organism to infer its own existence. This has been proposed by Fred Keijzer, Peter Godfrey-Smith, and others as the origin of subjectivity: having not only a point of view but also the experience of being a self, experiencing the world. Here, we can start to see the first glimmers of self-awareness that will ultimately be so important in understanding free will in humans, as we explore in later chapters.

But our next step is to see how all this subjectively extracted meaning is integrated with systems of planning and evaluation to inform action selection in increasingly sophisticated ways across evolution.

# 6

# Choosing

The elaboration of the neocortex allowed more sophisticated perception aimed at extracting meaningful information—identifying objects of interest in the world and creating a map of where they are and how they are moving, all relative to the organism itself. Perception became more and more internalized as additional cortical areas evolved and were recruited to the task of extracting higher- and higher-order features. It also became more subjective, in the sense of being tuned by prior experience and top-down expectations, allowing organisms to internally represent beliefs about what is out in the world. All of that neural machinery is expensive to build and to operate. Those costs are worthwhile only because that information could be profitably put to good use in guiding behavior.

Systems for action selection thus increased in sophistication in parallel with those for perception, with each improvement on either side creating new opportunities that could be exploited and further elaborated. Vision and hearing both can be used to map distant objects with respect to each other, creating an internal model of those spatial relations and how they are changing through time. This makes it possible for an organism to assess the entire situation in which it finds itself, in terms of the layout of multiple threats and opportunities. And it becomes especially valuable to learn about the *causal relations* between different objects or between the objects and the organism itself. In a world filled with other creatures, knowing which ones you can eat or mate with or should avoid makes the difference between life and death.

That is why memory systems are a really good investment, enabling an organism to retain knowledge of the past to inform current action. At the same time, the long-range nature of vision and hearing—literally, the ability to see something coming from a mile away—also made a new cognitive activity worthwhile: *planning*. The knowledge of distant threats and opportunities could inform the choice of longer-term goals and multistep strategies to achieve them. Action selection moved from very immediate responses to local stimuli to planning over much longer timescales. As with the evolution of perception, this capability relied on additional levels of internal processing, which were decoupled from the immediacies of the environment. The imagined future thus also came to crucially inform decision making as much as the remembered past. The organisms that evolved these capabilities no longer just inhabited the here and now.

These new neural systems for action selection did not replace or supplant the older ones but integrated with them as additional levels of control. To see how these elaborations evolved, we should look back at the control systems in simpler organisms, starting with the tectum, which was introduced in the last chapter.

## Coordinating Perception and Action

The tectum, located at the roof of the midbrain, is designed to map incoming sensory information onto a set of possible actions. We have seen how it can link specific types of visual signals to specific behaviors—for example, big, looming stimuli (possible predators) to avoidance, and small stimuli (possible prey) to orienting responses and approach (see Figure 6.1 for how this linkage works in the lamprey). But to direct more complex navigation of the environment, it has to do more than that.

First, it integrates information from multiple different senses—vision and hearing, most commonly, but also electrosensation and thermosensation in organisms like electric fishes or rattlesnakes that detect objects through these modalities. All these inputs are conveyed to the tectum, and the resultant maps are functionally aligned with each other, enabling the creature to derive a single integrative map of objects in the world. Second, the circuitry of the tectum performs various kinds of

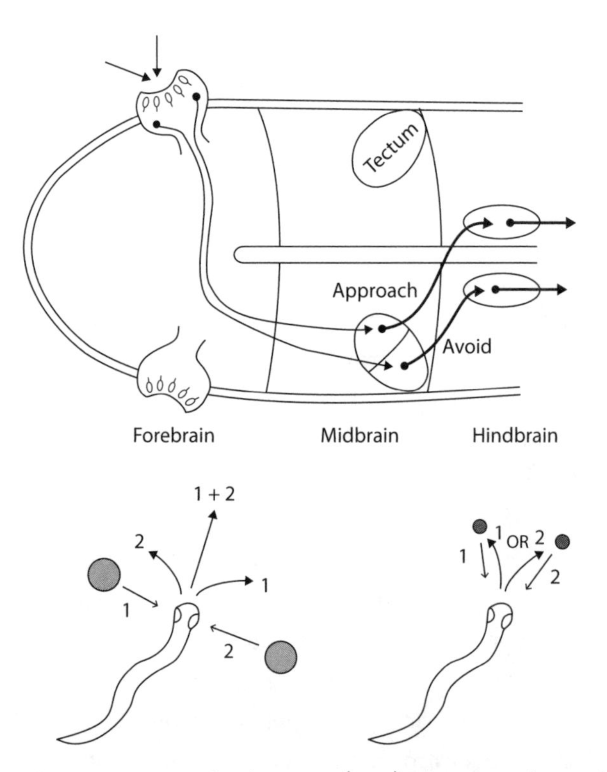


FIGURE 6.1. Action control in lamprey. **(Top)** The relatively simple visual system of the lamprey is linked to circuits in the tectum that mediate approach or avoidance. Large, looming shadows from the side (possible predators) prompt avoidance behaviors, whereas small fast-moving objects in front of the animal (possible prey) prompt orientation and approach. **(Bottom left)** If faced with multiple large visual stimuli, an adaptive response is to average the signals and chart a course between them (1 + 2). **(Bottom right)** If there are two opportunities in the visual field, averaging would not produce an adaptive response, because then both opportunities will be missed. Instead, competitive dynamics in the tectum are used to choose among options (1 OR 2).

neural operations, using these integrated maps of different kinds of stimuli to derive an optimal action. Threats and opportunities rarely present in isolation, meaning the right decision has to take account of the whole situation.

For example, the optimal response to multiple signals of aversive stimuli coming from different parts of the world may be an average of the responses to these signals if they occurred in isolation. If there are threats on both the left and right sides, then moving forward without turning may be an appropriate response (speeding up might be a good idea too). The situation differs, however, when an animal is faced with multiple attractive stimuli, all of which it might want to approach. Charting a course between such possibly rewarding objects would be of no benefit: instead, the animal should choose one or the other direction. To accomplish this, the circuits that process these kinds of attractive signals in the tectum engage a different mechanism: a winner-takes-all competition. Incoming signals from two attractive stimuli may stimulate two sets of neurons that would drive movements in different directions. But before any signal is passed to the hindbrain and spinal cord, mutual inhibition between these sets of neurons ensures that one of these patterns is reinforced and the other suppressed.

The new structures that evolved in the cortex and the rest of the forebrain perform similar operations, just on more data, informed by more context, and over longer timescales. The outcome of these deliberations is still the selection of one action to be released while competing options remain suppressed. For many types of actions, these higher regions still output this decision via the tectum and the motor command centers in the hindbrain.

## Elaboration of Forebrain Structures

We saw already how the cortex expanded in mammalian evolution and took on a key role in perception: making sense of sensory stimuli. Other structures and circuits of the forebrain evolved in parallel, linking this perceptual information to more sophisticated planning and action selection. These structures include the *hippocampus*—named for its seahorse shape—which sits along the inner edge of the cortical sheet. It is crucial in learning and memory, the bases on which organisms can come to understand the world.

The forebrain also includes the thalamus, which we discussed in the last chapter: it is a conduit or switchboard, controlling information flow in and out of, as well as across, regions of the cortex. And then there is a set of structures collectively referred to as the *basal ganglia,* which sit below the cortex. (The word "ganglia" means a cluster of nerve cells, and they are called "basal" because they sit at the base of the forebrain.) These structures sit between the cortex and thalamus in an extended circuit loop (really a nested set of loops) that is crucial in mediating the evaluation and selection of actions and in learning from the outcomes of these actions. These decisions are informed by signals—carried by chemicals like dopamine and serotonin—coming from midbrain regions, which convey information about reward, punishment, surprise, and generally the value of action outcomes relative to expectations.

Collectively, these structures and circuits mediate the following elements of decision making: assessing the situation, including what is out in the world and current internal states; determining current needs and adjusting goals accordingly; thinking of possible actions to achieve one or more of those goals; simulating and evaluating the likely outcomes of such actions; using these valuations to inform and ultimately resolve the competition among possible actions; releasing the action thus chosen and inhibiting all others; monitoring performance and adjusting as needed; assessing the resultant situation and updating internal models of the world and of the organism's own states; assessing the value of the actual outcome of the action; and, finally, adjusting connection weights within the system to either reinforce or downgrade that action pattern the next time such a situation is encountered. Let's look at some of the details of how those interlocking processes are implemented.

## A World of Possibilities

We saw in the last chapter that perception is neither passive nor neutral. Quite the opposite, in fact: it is utterly utilitarian and goal oriented. The point of perception is to guide action. That requires *interpretation* of a scene: What are the objects in it, what can be done with them, what can they do to the organism, where are the threats and opportunities, and what is the situation that the organism finds itself in? Answering those questions cannot be achieved just by the parsing of sensory data. It requires knowledge: the organism has to know what it is looking at (see Figure 6.2).

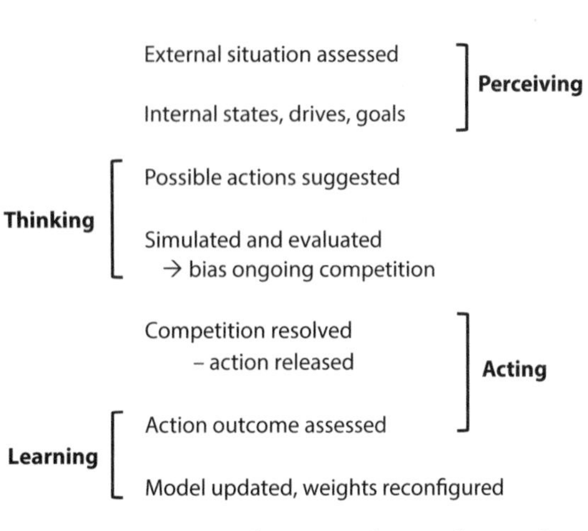


FIGURE 6.2. Action selection. Behavioral control involves an ongoing loop of perceiving, thinking, acting, and learning.

The circuitry of the cortex and hippocampus is tailor-made for accumulating such knowledge through learning and memory. At the simplest level, this involves learning the characteristic attributes of objects or regularly co-occurring stimuli, but the same principles apply to learning to recognize more complex scenes and remember past events. These processes rely on *synaptic plasticity*: processes at a cellular level that change the strength of connections between individual neurons.

When neurons that are interconnected are activated at the same time or in rapid succession, the strength of those connections tends to increase. That is, "neurons that fire together, wire together." That is how we learn that something out in the world—like a ringing bell, to use Pavlov's famous example—is associated with something else, like food. The learning in that particular example was actively motivated by a food reward, but more generally, cortical circuits are wired in such a way that they passively absorb the regularities of the things they encounter in the environment.

This is how we learn the properties of objects. We come to associate the color and form of a strawberry with its characteristic taste. Or the size and visual texture of a stone with an expected feel or weight. Or the shape of your mother's face with your concept of her as a person—

her personality, the kinds of things she might say, or the way she might react to something. The hierarchical layout of cortical circuits also makes them perfectly configured to learn about *types of objects,* which can themselves be organized into hierarchical categories. We can, for example, recognize our neighbor's dog Rex as an individual, but we can also think of it as a member of a particular breed, as a member of the dog species, as a mammal, and as an animal. And if we encounter a new animal with similar characteristics, we can similarly recognize it as a dog by connecting the current sensory percepts to our stored, high-level concept of "dogness."

The perceptual circuitry does not just segment the incoming sensory information into discrete objects—it connects to our memories to *recognize* those objects and activate a stored *schema* or constellation of the properties that different objects possess. We saw in previous chapters how incoming signals can be interpreted pragmatically through the context of configured control policies that embody "what to do about it." With the evolution of additional layers of processing, incoming data could be interpreted semantically in the context of stored knowledge. This connects percepts to concepts, enabling recognition and *understanding* of what is out there.

In turn, the remembered or inferred properties of an object inform the act of perception itself. We do not just perceive, say, the current visual aspect of an object; we perceive it as having a three-dimensional shape, even if we can't see it all, as having a back side or an inside, relative to our perspective. We subconsciously make predictions for what would happen and how our perceptions would change if we took certain actions; for instance, if we walked around the object, or touched it, or picked it up. These expectations rely on our knowledge of specific objects and types of objects and on our more general knowledge of the physics of our world. Even in the barest act of perception, you are creating and embedding that percept in a structure of possible actions and their perceptual consequences.

At the same time, we also are making inferences about other kinds of action opportunities and their consequences not for our perception of the object but relative to our goals. What can I do with that thing?

What can it do to me? The psychologist James Gibson called these properties "affordances," as in the answer to this question: What possibilities does this object or that object afford me? Some of those affordances are hardwired into the circuitry of simple organisms: a big looming shadow is *something that should be avoided.* A small object in front of you is *something that could be eaten.* But in animals with an expanded cortex, these affordances draw on our experience of how we interacted with different objects or types of objects in the past. A squirrel may know that a thin branch will not support its weight. A monkey may have learned that a stick can be used to extract ants from an anthill. A rat may believe it can drag a slice of pizza down the steps of a New York City subway.

What is important for the organism is to build up a picture of all the individual affordances in a given scene so it can assess the entire situation, with the same self-centered, utilitarian idea: What opportunities or threats does this scenario present *for me*? This requires a higher-order picture, considering objects in context and connecting that picture to knowledge we have built up from our experiences. The hippocampus is crucial in this process.

This forebrain structure is organized into several subregions, each with a distinctive layered structure. It receives inputs carrying information from most of the cortex, as well as other regions, and is functionally specialized to draw structural relations and contingencies from the convergence of different strands of information (see Figure 6.3). The neurons in the hippocampus are sparsely interconnected and tend to become active in small, transient groups, called *assemblies* or *ensembles.* Even when an animal is at rest, the hippocampus, like all areas of the brain, is intrinsically electrically active. Assemblies of coactivated neurons form and dissolve dynamically all the time, flowing from one such pattern to the next, with individual neurons transiently taking part in different groups. Importantly, these assemblies do not need to be driven by external inputs. Instead, a particular assembly can come to mean something *by being paired with* incoming stimuli during an experience. As the hippocampus integrates information arriving at the same time from

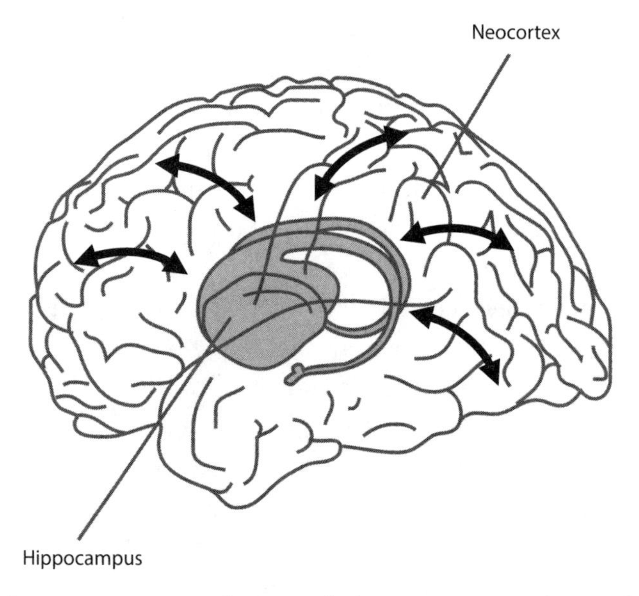


FIGURE 6.3. Hippocampus and cortex. The hippocampus sits deep within the brain at the medial edge of the cortex (on each side). It has extensive connections via the entorhinal cortex and other regions to all cortical areas.

different parts of the cortex, particular assemblies that happen to be active at a given moment can come to encode a memory of coincident stimuli.

This encoding seems to rely on the strengthening of connections within an assembly and between multiple assemblies that were activated at the same time. This makes it more likely that, in the future, when some of those neurons become electrically active, the rest are coactivated again, reinstating the patterns representing the triggering stimuli and thus activating a *memory*. This process begins with short-term biochemical changes within the relevant neurons, which alter things like the number of neurotransmitter receptor proteins on the surface of the cell, making one neuron more sensitive to signals from another one. With very strong or repeated stimulation, more stable changes arise, particularly the growth of additional synaptic contacts between specific pairs of neurons. This leads to the stabilization and consolidation of longer-term memories, which also require communication from the hippocampus back to the cortex.

In this way, the physical structure of neural circuits in the hippocampus and cortex literally comes to *embody knowledge*. These processes are analogous to—and indeed provide the inspiration for—changes in connection weights that underlie deep learning in artificial neural networks. This mechanism provides a bridge from the kinds of perceptual representations we discussed in the last chapter—patterns of concurrent neural activity that represent something out in the world at this moment—to representations of knowledge of the attributes of such objects, which is latently encoded in the pattern of synaptic weights within this network of neural circuits. The *meaning* of the neural activity patterns at any time is thus grounded by interpretation through the patterns of stored synaptic weights: incoming data acquire meaning by reference with stored knowledge. Perception thus entails more than just patterns being driven into the brain from the outside—it's the organism meeting the world halfway.

The hippocampus is specialized for making associations between pairs of stimuli and also for temporally ordering that information to form memories of *events*, which themselves are encoded in the hierarchical context of longer narrative sequences. These memories are framed relative to the animal itself—where it was, what it was doing, what happened next. At any moment, the hippocampus encodes a spatial map of the environment and where the animal is in it, which direction it is moving, and which way it is facing. As an animal moves, that picture changes, along with all the information about objects in the environment. The structure of the hippocampal circuitry is designed to link the assemblies active at one moment with those that are active in the next moment, and the next, and so on. This creates a temporally structured record of what happened during some episode: what is referred to as *episodic memory*.

That kind of memory is perfectly structured for drawing inferences about *causal relations*. If, in a given episode, A happened and then B happened, then maybe A caused B. If this relation were repeatedly observed over many episodes, then that causal inference would become better justified. Moreover, the hierarchical relational structure of different memory elements allows organisms to build up a complex picture of causation through the binding of items and events within a contextual framework. It may be that A leads to B but only in the context of C—this too is a regular relation that the hippocampus and cortex are configured to recognize.

Of course, we don't want to remember everything that happens to us. Forming memories is energetically expensive: resources are required to build and maintain those new synapses. In addition, a lot of things that happen are spurious and do not represent reliable regularities in the world. Or they may be of little relevance to us and better ignored. Reflecting this, the cellular processes of synaptic plasticity do not happen by default: they are gated by processes of attention, arousal, salience, reward, or surprise and signaled by neuromodulators like dopamine, serotonin, and acetylcholine.

This entire system is finely crafted to learn just what it needs to learn to create a store of knowledge that the organism brings to bear in assessing any new situation (see Figure 6.4). That includes the knowledge of past events, especially the actions that the organism itself took and their outcomes. With the evolution of visual perception, we saw the need for the organism to take its own movements into account. This applies to the causal model of the world too: making sense of all these relations requires *modeling the self as a causal agent*. The organism has to know, either implicitly or explicitly, that it can cause things, that its own actions have predictable outcomes. In turn, it acquires causal knowledge through its own history of intervening on the world and observing the results.

The perception of any animal is thus embedded in a field of imagined scenarios or actions and the sensorimotor consequences that would be associated with them. It is literally *a world of possibilities*. Even in the barest act of perception, you are creating and embedding that percept in a structure of possible actions to answer the question: *What can I do here?* But to actually act, to choose from all the possible options, an organism also needs to answer the question: what *should* I do here?

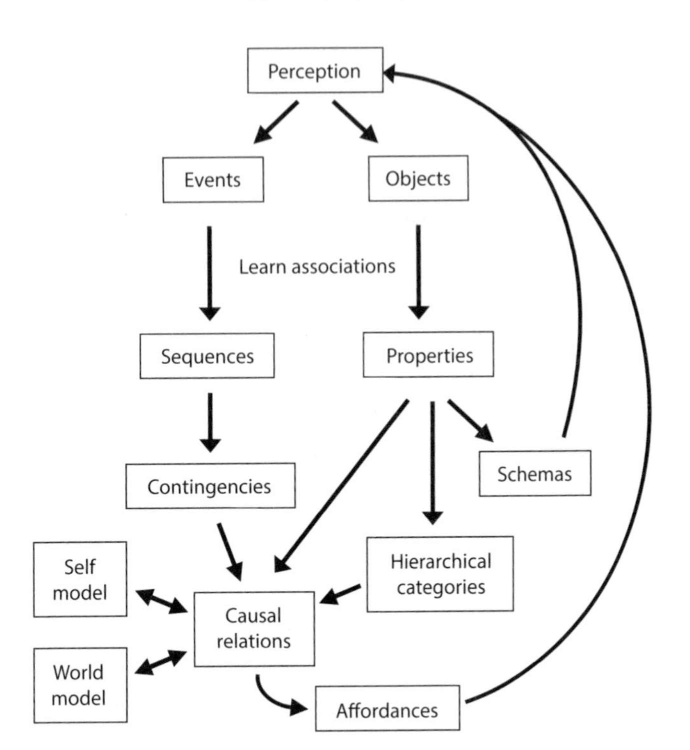


FIGURE 6.4. Acquiring knowledge. Hippocampal and cortical circuits are adapted for learning associations, including properties of objects and sequences of events, and for abstracting categorical relationships, schemas, contingencies, and causal relations. These relations come to be embodied in the configuration of neural circuits, thus comprising a web of knowledge that underpins adaptive behavior. This knowledge is brought to bear on the act of perception itself, which is an active, egocentric process aimed at inferring what the organism can do in any given situation.

## What's My Motivation?

What an animal should do depends not just on the external situation but also on its current needs, which depend on the internal state of the animal. All kinds of bodily parameters—energy balance and nutrient status, fluid balance, sleep need, and body temperature, for example—have to be kept within tight operating ranges for the animal to survive. In most vertebrates, these parameters are monitored by the hypothalamus in concert with connected regions in the midbrain. Any deviation from an optimal set point initiates a counteracting response, either physiological or behavioral.

The hypothalamus is a key nexus between body and brain, monitoring signals in the bloodstream and neural signals coming from multiple parts of the body. It also has dual modes of action in the opposite direction, controlling the release of many types of hormones back into the bloodstream and directing autonomic nervous system signaling. These processes act in concert to regulate bodily functions—blood pressure, heart rate, enzyme levels, digestion, gut peristalsis, sweating, shivering, sleep, arousal, and others—in ways that aim to bring essential parameters back into line.

With the same aim, information on current needs is also used to prioritize certain behaviors. The hypothalamus is extensively interconnected with other brain regions that collectively provide signals that motivate different types of behaviors, with varying levels of urgency. The resultant signals of need may be experienced as hunger, thirst, tiredness, cold, pain, fear, and so on; that is, basic emotions. These naturally lead to urges to eat, drink, sleep, find shelter, avoid whatever is causing pain or fear, or respond with defensive or aggressive behaviors. All these are clearly essential to the survival of the animal.

These are the most immediate and proximal needs of the animal, and they powerfully influence behavior. But other imperatives are equally important over longer timeframes. The need to reproduce, for example, and the drive to engage in mating and parenting behaviors are also controlled by circuits in the hypothalamus. In social animals like humans, social behaviors—staying with a herd or family group, maintaining social relationships, achieving more dominant status, and so on—are just as essential to both survival and reproduction. More complex emotions like loneliness, jealousy, and anger can thus also be strong drivers of behavior in many species. And in many situations, the right thing to do is to wait for or to actively seek more information, and that can become a goal as well.

The overall profile of current needs is conveyed by the hypothalamus and connected regions to the parts of the brain involved in selecting actions. In lower vertebrates like the lamprey, for which behavioral options are limited and cognition is relatively shallow, these parts of the brain prominently include the tectum and midbrain regions. In mammals, which have a greater range of behaviors to choose from and greater cognitive depth, this information is also conveyed to the cortex and other parts of the forebrain. These signals help determine what type of behavior the animal should engage in. But they don't direct the specific actions that should be taken to mediate that behavior—those are context specific.

For example, if an animal has low nutrient status, it will send a hunger signal to the decision-making regions. How this is acted on depends on the situation. If there is food available, the animal should eat it. If there is no food, the animal should seek it. Doing so may involve all kinds of behaviors—crying, begging, trading, stealing, grazing, foraging, scavenging, hunting—and each of those activities involves a variety of possible actions. The same principles apply for other behaviors, which can be divided into *consummatory* or *appetitive* components: if you've got what you need, then complete the action (feeding, drinking, mating, sleeping); if you don't have what you need (food, water, a willing member of the opposite sex, a safe place to sleep), then seek it out. Essentially, the animal makes a decision to exploit or explore, depending on what is currently directly available. The signals from the hypothalamus convey the relative urgency of these needs—broadly speaking, what to do—but not how to do it.

These signals feed into the overall decision-making processes of the animal. They may be especially important in supplying the necessary drive to justify the investment of energy and time and risk that the animal must commit to perform various actions. A very hungry animal will prioritize seeking food over other activities and will also work harder and take more risks for food than a satiated one. Of course, the effort expended by the animal to find food must be balanced against other current and even anticipated needs and opportunity costs. The relative level of all these signals thus informs the goals that an animal chooses to pursue at any one time.

In most mammals, those goals are quite simple and immediate; they are closely related to their underlying needs. We will see in later chapters how humans can build a hierarchy of nested goals that stretch over longer timeframes and that require more complex behaviors to achieve. Goals are thought to be represented in the cortex, likely in a hierarchical fashion proceeding from immediate, short-term goals (or action plans), which may be represented in the motor cortex, to longer-term goals that rely on more frontal regions, known as the premotor and prefrontal cortex. We will return to the prefrontal cortex in later chapters.

## Simulating Possible Futures

So, now we have an animal that has surveyed the scene, compared it to stored memory, recognized and mapped objects, characterized and assessed the entire situation, and taken account of its own internal states to prioritize certain goals. The next task is to integrate all this information so it can decide what specific action to take. In some circumstances, there may be one clear goal and one obvious action to take to achieve it. But most of the time there are multiple goals in play, possibly conflicting with each other, at least in the sense that pursuing one may cost the opportunity to pursue the others. And there may be many possible actions, all of which are viable—indeed, too many to efficiently deliberate between. There may be too many even to conceive of them all.

Imagine a monkey that is not starving nor dying of thirst nor facing any imminent threats but is reasonably safe in the company of its extended troop. There is a huge range of things it could do. It could climb a tree, search for grubs, scratch itself, groom another monkey, look for a mate, go to sleep, bang some rocks together, jump up and down, poke itself in the eye, urinate, start a fight, wave its arms around, stick a pebble up its nose, eat some dirt, screech, and on and on and on. But it doesn't actually have to deliberate between all these actions because most of them do not "spring to mind": they just do not occur to the monkey as possibilities in that moment.

What controls which possibilities do spring to its mind is a bit of a mystery. This seems to rely on past experience but in a quite general way, one not so much related to the specific context. Our monkey may know that searching for grubs and grooming another monkey and going to sleep *generally* are all profitable ways to spend its time. These activities may thus be the ones that occur to it, while it doesn't even consider the vast array of other possible actions. You may have noticed that some of the possible actions on that list, like screeching or eating dirt or sticking a pebble up its nose, are things that infants—monkeys and humans alike—do actually engage in. They explore all kinds of options, without much apparent discrimination, but they learn over time which ones tend to pay off and which ones do not.

In other words, animals develop, through experience, *habits of thought*. We will consider such habits in more detail when we discuss free will in humans in later chapters—especially the notion that if such ideas just spring to mind, then we are not really in control of them. But for the moment, the important thing is that this shortcut dramatically narrows the search space of possible actions.

The next step is to choose among those actions. In theory, an animal could just try different ones out and see what happens. The trouble with that trial-and-error approach in the real world is that "error" often implies death. At the very least, in a competitive world, it means losing out to rivals on possible gains and opportunities by spending time and effort on actions with low payouts. A far better strategy is to internally simulate a set of possible actions and predict and evaluate their likely outcomes. These actions can then be weighed against each other to enable selection of the one with the highest utility relative to the organism's current goals. There is an obvious benefit to evaluating potential actions offline without actually engaging with the world, providing it can be done with enough efficiency that the organism is not caught dithering.

These processes are similar to those we engage in when we play a game of chess. At any point in the game, there may be ten or more pieces that we could move. When selecting among these options, we go through in our minds what would happen if we moved the queen to this square or that square or if we moved the bishop or the knight or this pawn or that piece. We simulate in our minds the consequences of each possible move and evaluate whether they are good or bad. In some circumstances, the right move is obvious; in others, it requires looking further and further into the future, along more and more branching paths of moves and countermoves. Some moves may incur a short-term cost but open up a longer-term opportunity. Some would likely be so costly that they would cost us the game, if not immediately, then in short order.

The point of simulating all these options in our minds is that we don't have to try them out and incur that risk in the game itself. The benefits of this approach in life, where the stakes are much higher, are obvious. The philosopher Karl Popper famously said that this capacity for mental simulation allows us to "let our hypotheses die in our stead." And, of course, the more experience we gain—in life, as in chess—the better we are at making these decisions, recognizing situations, narrowing our search space of options, predicting outcomes, and accurately evaluating them over a longer horizon into the future.

In vertebrates, these processes of simulation, evaluation, and eventual choice are mediated by an extended set of nested circuit loops between the cortex, basal ganglia, and thalamus, with inputs from midbrain centers carrying utility signals and outputs to motor command centers in the tectum and other parts of the midbrain. The exact details of what all these elements do is a matter of active debate and research, but a broad sketch of our current understanding is as follows.

First, some action plans are conceived of in the cortex. Doing so may entail the activation of different sets of neurons, with each specific pattern corresponding to a particular action plan. At this point, these patterns represent *the idea of* doing something, not a commitment to it. These cortical patterns of activity are conveyed through a massive fan of parallel nerve fibers to the input region of the basal ganglia, called the *striatum* (so named because this fan of incoming fibers gives it a striped or striated appearance). Neurons in this region are hard to activate. By contrast, the output regions of the basal ganglia (called the GPi and SNr) contain inhibitory neurons that are active nearly all the time. They project to the midbrain motor centers and keep the brakes on all the motor command neurons in those regions. The baseline function of the basal ganglia is thus to *inhibit all actions*.

Signals from the striatum can override this inhibition to allow some activity to occur, but they do so in a complicated way. The striatum has two types of neurons, which are both inhibitory and which project through two different pathways to the GPi. One set projects directly to the GPi and inhibits GPi neurons in a pattern-specific way, thus *disinhibiting* the motor command centers that correspond to the intended action. The other set of striatal neurons does the opposite. They project indirectly to the GPi via another subpart of the basal ganglia called the GPe. GPe neurons are also inhibitory, so we end up with a triple negative loop (inhibiting the inhibitors of the inhibitors), which effectively activates the GPi neurons, reinforcing their own inhibitory control over motor centers.

What is the point of this complicated circuitry? One (admittedly controversial) way to think of these two pathways is as representing the expected *costs and benefits* of each action plan. If the costs are high, the indirect ("no-go" pathway) will be very active. If the benefits are high, the direct ("go" pathway) will be more active. The relative merits of each plan can then be compared to select the best one. The various action plans are in a sense competing with each other to take the wheel at any given moment. The function of the basal ganglia in this loop may be to bias the ongoing competition between the patterns of activity representing the various action plans in the cortex itself, giving some of them a boost relative to the others. This function relies on a recurrent loop that runs back from the basal ganglia to the thalamus and cortex (see Figure 6.5). There thus seems to be an internal loop for *thinking about what to do* and an output circuit for actually releasing the chosen action, while still inhibiting all the others, once all the thinking is completed. (To further complicate matters, there is another pathway called the hyperdirect pathway that bypasses the striatum altogether and lets the cortex very rapidly veto or terminate actions.)

One of the functions of the basal ganglia in this process is to help integrate current motivational and situational information with sensorimotor action plans. Neuromodulatory circuits convey the motivational signals discussed earlier to the striatum, where they regulate what engineers call the *gain* of different channels, effectively energizing the

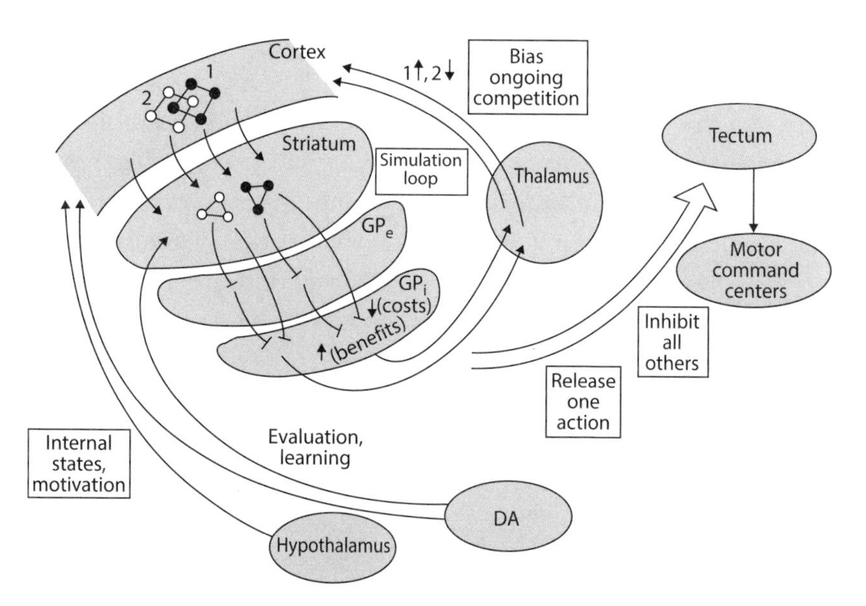


FIGURE 6.5. Basal ganglia–cortical loops. In any given situation, multiple possible goals or actions may be primed as neural patterns in various cortical areas. At any one level of the nested hierarchy of behavioral control (for goals, plans, actions, or movements), these patterns are effectively in competition with each other for control of the actual machinery of movement. Transfer of the patterns through the basal ganglia allows an internal evaluation of their merits, based on prior learning and current states, informed by signals from dopaminergic midbrain nuclei. The parallel inhibitory and activating (really disinhibiting) circuits through the globus pallidus may effectively assess costs versus benefits for each action, with ones that come out on top in that calculus receiving a boost in the ongoing competition, via signals back to the cortex from the thalamus. Activity in this *simulation loop* continues until one action plan emerges as a clear winner, at which point signals conveyed to the tectum and the midbrain command centers release and invigorate that action while keeping all others inhibited. Monitoring of the subsequent outcome then instructs learning by modifying synaptic weights representing action–outcome associations in the striatum and cortex, thus allowing the organism to learn from experience.

choice of some actions over others. These circuits use neuromodulators like dopamine, which in this context do not cause the recipient neurons to fire on their own but do make them more sensitive to activation by the incoming cortical axons. If the animal is hungry, then incoming patterns of activity that represent plans for seeking food are amplified, reflecting the fact that the benefits of finding food are greater in the hungry state (or that the costs are more worth bearing). These signals may also regulate the vigor with which a selected action is executed.

Dopamine and other neuromodulatory signals (including serotonin, acetylcholine, noradrenaline, and histamine) are also involved in the selection of actions based on their predicted outcomes. Predictions are made in the cortex and hippocampus, based on the learned regularities of the world and the knowledge gained by past experience with sequences of events and causal relations within them. This knowledge is, in a sense, remembering the future. These processes generate predictions of rewards or punishments and the overall subjective utility of the predicted outcomes of various actions that the animal uses to prioritize actions and that it also monitors in an ongoing way as actions are being executed. Deviation from these predictions induces dopamine signals from the midbrain, which have strong behavioral salience; in effect, they tell the animal that things are not working out as predicted and that a change of action may be a good idea.

Note that there isn't anything special about these neuromodulatory chemicals themselves. There is nothing inherently rewarding or surprising about dopamine, for example. What its release means depends entirely on the context: what stimuli are represented by the dopamine-releasing neurons and how that information is interpreted by the receiving circuits. In these scenarios, these neuromodulators are not widely released and do not act like global hormones: the brain is not bathed in a certain level of dopamine or serotonin. These molecules are released very locally under tight control at specific synapses within these circuits. In the context of action selection, dopamine signals deviation from predictions, good or bad (which you could call surprise or disappointment or delight). It is released from neurons that compare behavioral outcomes to internal predictions, and those signals are interpreted in the striatum and elsewhere by up- or downregulating the gain of various channels, allowing the animal to guide ongoing action.

Perhaps the most important function of these circuits, however, is enabling the system to learn from its experiences so it can make better decisions in the future.

## Teachable Moments

The process of behavioral control is not a series of static, unrelated events. The organism does not have to make every decision from first principles. Instead, it brings what it has learned from its own history to bear in every new moment of decision making. It does so by keeping a record of how its past choices turned out and reinforcing the ones that turned out well.

Again, dopamine is a central player in this process, particularly in evaluating the outcomes of actions. A leading model for how this works goes like this: when a certain action turns out well, dopamine signals lead to increased synaptic plasticity at synaptic connections from the cortex to the striatum and, via the return influence of the basal ganglia, within the cortex itself. These processes tend to strengthen the connections for actions that had good outcomes and weaken those for ones that turned out poorly. The effect of this kind of instrumental or reinforcement learning is that the next time the animal is in that situation, or one like it, that reinforced action will automatically have a leg up in the competition. The incoming cortical signals will impart a stronger signal to the striatal neurons, automatically upregulating the activity representing the benefits of that action relative to all the others.

This kind of reinforcement learning is a key process in deep learning in some artificial neural networks. It allows them to be trained to complete specific tasks by giving them feedback on how well they did the last time, which then is used to adjust connection weights in the network that represent different ways of manipulating the input data. The key element of both artificial networks and the brain is that the record of past performance is not stored in a separate region that would then have to be consulted every time a decision has to be made; instead, the learning is baked right into the decision-making machinery itself. (That is not to say that we don't also make explicit episodic memories of many events, but this process is not necessary for this kind of reinforcement learning to occur.)

Each time the organism chooses an action, there is an opportunity for reinforcement learning to occur: each becomes a teachable moment.

The more salient the outcome, the stronger the resultant changes in neural connections and the bigger the effect on subsequent behavior. Over time, if an action turns out to be a reliably positive thing to do, it may become so strongly reinforced that it becomes a habit—both of thought and of execution. In fact, it may become so automatic that simply the recognition of a situation may trigger the behavior, without any conscious thought having to be given to the goal or to other options for behavior in that context.

We tend to think of habits as bad things, but really they are tremendously useful shortcuts that enable animals, including us, to navigate familiar settings and scenarios in adaptive ways with a minimum of cognitive effort and time expended on deliberation. We've done all the hard work of thinking about this already, so why do it again? Our brains know how things are going to turn out, broadly speaking, if we behave in tried-and-true ways in most of our everyday contexts. People are sometimes asked, "If you could go back in time and give your younger self some advice, what would it be?" In reality, what happens is precisely the opposite: our past selves are giving advice to our present self all the time to ensure it has the best possible future.

To summarize, these extended brain systems, involving circuits within and between the cortex, hippocampus, hypothalamus, thalamus, basal ganglia, midbrain nuclei, motor command centers, and other regions, collectively mediate these diverse and integrated processes of behavioral control. These processes enable the animal to identify and assess the situation, monitor current needs and prioritize different goals, conceive of possible actions, select among them, and learn from their outcomes to inform the process in the future. These capabilities let animals further decouple perception from obligatory action, test out possibilities in their heads before risking them in the real world, and recruit their whole historical selves in the service of making optimal decisions.

Along our own lineage, the progressive expansion of the forebrain, in particular of the cerebral cortex, allowed animals like to us to increasingly internalize these processes—generating additional levels of the hierarchy for learning and planning over longer timeframes. As we will see in later chapters, this eventually resulted in the capabilities of introspection and metacognition, through which our own goals and drives and ideas become not just the elements but also the objects of cognition. In this way, mammals, primates, and eventually humans developed more and more layers of insulation from the exigencies of the environment and greater and greater autonomy as causal agents in their own right.

## Neither Ghost Nor Machine

There are several important points to emphasize regarding the picture presented in this chapter. First, organisms do not passively wait for external stimuli to respond to. Their brains, when awake, are constantly cycling through possible actions, and this stream of behavior *accommodates to* new information and the changing environment. Second, this is not a one-way relationship from environment to organism: it is a recursive loop of mutual interaction. The activity of the organism changes the environment and the organism's relation to it. The apparently linear chain of causation is really a loop or a series of loops—you can think of it as a spiral stretched through time. If we ignore these reciprocal effects, we are left studying only half the overall system. Third, the processes of decision making and action selection are just that—*processes*: they have duration through time. They are not instantaneous transitions from one physical state of the system to the next. This point is crucial when we consider some philosophical challenges to the idea that choices can be made at all.

Finally, the description of the processes involved in action selection risks giving the impression of a mechanism churning away or of a computer running a linear algorithm. Of course, there are mechanisms at work: we can see this clearly when parts of the brain are damaged or altered by drugs, and decision making goes awry. And it is true that some of the operations of these mechanisms can be thought of as computations. However, the idea of an algorithm—a series of steps being completed methodically and sequentially—is not an accurate conception of what is happening. The various subsystems involved are in constant dialogue with each other, each attempting to satisfy its own constraints in the context of the dynamically changing information it receives from all the interconnected areas. Ultimately through these dynamic, distributed, and recursive interactions, the whole system settles into a new state—one that drives the release of one of the set of possible actions under consideration and the inhibition of all the others.

In a holistic sense, the organism's neural circuits are not deciding—*the organism* is deciding. It's not a machine computing inputs to produce outputs. It's an integrated self deciding what to do, based on its own reasons. Those reasons are derived from the meaning of all the various kinds of information that the organism has at hand, which is grounded in its past experience and used to imagine possible futures. The process relies on physical mechanisms but it's not correct to think it can be *reduced to* those mechanisms. What the system is doing should not be identified with how the system is doing it. Those mechanisms collectively comprise a self, and it's the self that decides. If we break them apart, even conceptually, we lose sight of the thing we're trying to explain.

However, although we can reject a reductionist, purely mechanistic approach, that should not send us running in the other direction toward a nebulous, mysteriously nonphysical mind that is "in charge": the ghost in the machine. Our minds are not an extra layer sitting above our physical brains, somehow directing the flow of electrical activity. The activity distributed across all those neural circuits produces or *entails* our mental experience (and similarly for whatever kinds of mental experience other animals have). The meaning of those patterns for the organism has causal power based on how the system is physically configured. We can thus build a holistic physical conception of agency without either reducing it or mystifying it.

# 7

# The Future Is Not Written

The story I told so far charts the emergence in the universe of causal agents, from the origins of life in the rocky crevices of the ocean floor, to the invention of behavior, and the evolution of more and more complex systems for perception, action, and decision making. Through the cumulative effects of natural selection across millennia and of learning over the course of individual lifetimes, living things accrete causal power. Although they are made of physical components, they are not *merely* physical systems, where the things that happen within them are driven by low-level causes. They are organized for a purpose, and that organization constrains the physical components to enable true functionality and goal-directed action. Their physical structures are configured so that they run on meaning, on patterns of activity that *represent* things—percepts, concepts, beliefs, needs, goals, plans, causal relations, regularities of the world, memories, scenes, narrative sequences, and possibilities.

But what if all that is an illusion? What if living organisms—including us—are not really choosing anything at all? What if there are no possibilities in the first place?

## Is the Future Written?

In the first chapter, I introduced two flavors of determinism ("hard" and "soft") that present overlapping but distinct challenges to the philosophical idea of free will in humans. To avoid conflating them, I will call them *physical predeterminism* (the idea that only one possible timeline exists) and *causal determinism* (the idea that every event is necessarily caused by preceding events—usually seen as the same thing as physical predeterminism but subtly distinct). And I will add a third flavor that we will need to tackle too: *biological determinism* (the idea that an organism's apparent choices are really internally necessitated by its own physical configuration: its biochemical state or nervous system wiring).

I deferred discussion of these challenges until now but tackle the first two types of determinism head-on in this chapter. These ideas are not just problems for free will in humans: they present an equally stern challenge to the question of whether any living organisms can really be said to have agency at all. We will see that the real issues stem not only from deterministic thinking but also from a *reductionist* viewpoint, especially the idea that all the causal influences arise at the lowest levels of reality: that of subatomic particles and fundamental physical forces.

The idea of physical predeterminism is both ancient and widespread. The Greek philosopher Democritus, who first proposed the existence of atoms—the tiniest indivisible units of matter—also argued that their trajectories through time do not diverge from predestined linear paths. Fast-forward two thousand years, and Newton's laws of motion seemed to fully support this view. These laws of what is now known as "classical" mechanics are (supposedly) completely deterministic. If you know the position and momentum of every component of a system (for simple systems at least), these laws can be used to predict the next state of the system with high degrees of accuracy—enough, for example, to predict solar eclipses hundreds of years in the future.

Several hundred years later, Einstein's work on relativity also seemed most consistent with physical predeterminism. He saw time not as distinct from space but as a fourth dimension of space-time, building up a picture of a static block universe in which all space and time points are essentially given at once. In this picture, the future is as determined as the past. We happen to occupy a certain slice of space-time at any given moment, but the trajectory of how the universe unfolds is already fixed. Note that, in this model, it is not clear what determines *which moment* we are subjectively experiencing, because nothing objectively distinguishes the present from the past or future. But the important point is that the state we find ourselves in "right now," whenever that is, emerged completely deterministically from the prior state and completely deterministically predicts the next state. In fact, in the block universe, time has no inherent directionality: nothing in that formulation states that it is more likely to flow toward the future than toward the past.

The same deterministic scenario was described in 1814 in a famous thought experiment by the French polymath Pierre-Simon Laplace. He imagined an omniscient intellect (now known as Laplace's Demon), powerful enough to apprehend all the details of the state of the universe in an instant and, in doing so, to grasp immediately the universe's entire trajectory of existence, both forward and backward in time:

> We may regard the present state of the universe as the effect of its past and the cause of its future. An intellect which at a certain moment would know all forces that set nature in motion, and all positions of all items of which nature is composed, if this intellect were also vast enough to submit these data to analysis, it would embrace in a single formula the movements of the greatest bodies of the universe and those of the tiniest atom; for such an intellect nothing would be uncertain and the future just like the past would be present before its eyes.[1]

This same view is summed up in the definition of determinism given by philosophers Gregg Caruso and Daniel Dennett in a 2021 book debating the subject of free will and moral responsibility: "Determinism: The thesis that facts about the remote past in conjunction with laws of nature entail that there is only one unique future."[2] Notably, both seem to agree that this thesis is true, though they diverge markedly on its implications.

1. Pierre Simon Laplace, *A philosophical essay on probabilities* (New York: Wiley & Sons, 1902), 4.

2. Daniel C. Dennett and Greg D. Caruso, *Just deserts: Debating free will* (Cambridge: Polity, 2021), 5.

## The Lowest Levels of Reality

This position fits with one interpretation of what is known as the *core theory* in modern physics. This theory encompasses the so-called Standard Model of particle physics, along with the theory of general relativity, within certain limits. The physicist Sean Carroll describes how the core theory effectively covers all the ranges of physical interactions that arise in the domain of normal human experience. (It may break down at extremely high energies or near a black hole, but those exceptions don't bear on its applicability in the world we actually experience.)

The core theory deals with the smallest elements of our universe and the most elementary physical forces. It is a theory of *quantum fields* and how their interactions play out through time. We are used to thinking of matter as being made of particles—atoms in the first instance, which are themselves composed of subatomic particles: electrons, protons, and neutrons. These combine in different numbers to give all the chemical elements, with their respective properties. It turns out that protons and neutrons are themselves made of even smaller particles called up and down quarks. Electrons and up and down quarks, along with one more particle—the enigmatic neutrinos, which do not interact much with anything—comprise the basic building blocks of matter.

In addition, there are four basic forces—electromagnetism and gravity, both of which are familiar from everyday life, as well as the strong and weak nuclear forces that act at the level of these subatomic particles and, along with electromagnetism, explain how they stick together in protons and neutrons and atoms in the ways that they do. All these forces are also mediated by various particles (like photons for the electromagnetic force, for example). These *force particles* (technically known as bosons) are different from the *matter particles* (known as fermions) in that they do not have mass or occupy space in the same way—although it is impossible to put two electrons in the same spot, you can pile as many photons into a piece of space as you like.

This picture of matter and force particles is not the whole story, however. Quantum physics has revealed that these particles can have wavelike properties as well, as if they can be spread out through space somehow. Indeed, a better way of thinking of them is as a distributed quantum field that sometimes shows discontinuities that we can think of as similar to condensation into a particle. One of the key findings of quantum mechanics is that these condensations only happen at discrete energies. The field is thus not continuous with respect to where the particles might be or what their properties might be; instead they are *quantized* into specific possible states that they can sometimes "jump" between.

For our purposes, what is important is that the laws and equations governing how these basic particles and fields behave within the regimes that we may ever experience seem to be fully known. It is not that there are no other physics required to understand things on a cosmic scale or at enormous energies; indeed, there must be. But at the level of our own existence, these equations make predictions (about things like the energy of various particles) that, as shown in experiments, are extraordinarily accurate, with measurements of predicted parameters confirmed to twelve or thirteen decimal places in some instances. Moreover, it can be argued on theoretical grounds that these equations are *causally comprehensive*: not only do they explain how quantum fields behave and evolve through time but they also rule out the existence of any possible additional causal factors. There is no need for anything else and *no room for* anything else.

That at least is the claim. It implies that for any system under study, if you have all the details of the particles and fields at "time $t$," they can in theory be plugged into the equations, which will tell you what the system will be like at "time $t + 1$." And the same for the next time step, and the next, and the next, and so on, forever. That claim clearly poses problems for the ideas of agency and free will. If the brain is a physical system, made of the same basic elements of matter as everything else in the universe, then the laws of physics should control how it evolves through time and will do so deterministically. And if the physical states of the brain entail your conscious experience and your behavior at any moment, then those are just as determined by how these physical interactions play out at the lowest levels of reality. Indeed, the behavior of everything in the universe would have been *predetermined* from the dawn of time, including me writing this sentence and you reading it.

In my view, this claim is absurd on its face. That is not an argument for whether it's true or not but just an observation that this claim jars so bracingly with our actual experience of the world that we should be strongly suspicious of it. (And, indeed, we will see later that it is undermined by quantum physics itself.) But if we take it as a premise, it is clearly a problem not only for free will as variously conceived in humans but also for the more fundamental concept of agency itself. How can you say an organism is *doing something*—that *it itself* is a cause—if what's happening is just the inevitable manifestation of the interactions of physical particles within it? That claim certainly doesn't leave any room for the organism to *choose* what to do. In that kind of predeterministic scenario, there is nothing to choose between. There is only one future open, with no possibilities, no deciding or acting, no purposiveness, no mattering, no trying, no goals, or functions—it would be literally a meaningless universe.

## Escaping with Our Morality Intact

The picture I just drew is pretty stark and difficult to square with the existence of free will. But there are arguments to be made that some kind of free will is indeed compatible with this kind of physical predeterminism. Indeed, some version of this kind of *compatibilism* is probably the most popular view of scientists and philosophers on the matter. One of the most prominent proponents of this way of thinking is Daniel Dennett, who in many philosophical papers and popular books (including the one just mentioned) presents a comprehensive framework that argues for the existence of what he calls a kind of "free will worth wanting."

Retaining some semblance of free will is motivated here by the desire to retain some philosophical foundation for the concept of moral responsibility. If the future is fixed and all events just flow according to the low-level laws of physics, there seems to be no justification for saying that a person is responsible for his or her actions. In particular, if we can only hold someone accountable for an action *if they could have done otherwise* in a given situation, then this is ruled out by physical predeterminism. "Otherwise" doesn't apply in such a universe because nothing other than what was going to happen ever happens.

Dennett argues instead that what is important for moral responsibility is that individuals themselves were *the source* of the determining causes—that they did some act (let's call it A) *because they wanted to do* A. Had they wanted to do B, they would have done B. In fact, Dennett sees the question of whether they could have done otherwise if we "rewound the tape" and put them back in *precisely* the same physical conditions as nonsensical and irrelevant. He argues that of course they could not: the physical circumstances at any instant determine the next state of the system (of the whole universe, in fact), which necessarily encompasses the action that these individuals will do. What is more important for him is the ability to do different things if the conditions were slightly different. He argues that "the general capacity to respond flexibly . . . does not at all require that one could have done otherwise in . . . any particular case, but only that under some variations in the circumstances—the variations that matter—one would do otherwise."[3]

Dennett builds up a picture, not dissimilar to the one I laid out in the preceding chapters, for how evolution by natural selection can create organisms that have precisely this kind of causal autonomy and that act *for reasons*. This causal potential is accumulated across evolution through feedback processes that enable organisms to build information about their environment into the configuration of their own physical structures and use it to guide action in an adaptive fashion. Dennett further argues that the inherent complexity of the resultant structures—the product of billions of years of research and development by evolution—is so great that, even if all the mechanisms are still subject to the deterministic laws of physics, they remain entirely unpredictable in practice. This unpredictability supports a pragmatic view of seeing the causality as inherent *in the organism itself*, rather than in the machinery within it.

3. D. C. Dennett, I could not have done otherwise— so what? *Journal of Philosophy* 81, no. 10 (1984): 557.

The physicist Sean Carroll advances a similar view, which he calls "poetic naturalism."[4] He claims it is reasonable and appropriate to deal with different matters in the natural world at the level at which they manifest. Thus, we can develop effective theories within chemistry or biology or psychology or sociology that deal with the kinds of entities and processes that manifest at those various levels. We don't need to go down to the level of electrons and quarks and the forces acting on them to understand the dynamics of ecological or economic systems, for example. Indeed, it would be pointless to do so. Even if we could collect all the relevant data, we would be overwhelmed by meaningless low-level details and be unable to detect or comprehend the causally important patterns and principles that control how such systems at these various levels behave. It is much more convenient to winkingly attribute causality to persons, even though we know that in reality they are made of physical elements that are themselves entirely controlled by the low-level forces acting on them in a way that Carroll says is "causally comprehensive."[5]

These compatibilist ideas are certainly appealing, and perhaps they're correct—if so, then so much the better for the story I am telling in this book. If agency and free will can emerge even in a totally physically deterministic universe, then that is one less problem to solve. But, to my mind, these arguments are ultimately unsatisfying and unconvincing. They seem to argue that we can treat humans and other organisms *as if* they have some causal power, even though we know that they are not really making choices. It is only because they are so complicated that, for all practical purposes, we can and should treat them as reasoning agents with the capacity to act, and not just as complex arenas in which the laws of physics play out. However, these arguments do not resolve the central problem raised by physical predeterminism: they just tell us not to worry about it, that it's not the problem we thought it was.

4. Sean Carroll, *The big picture: On the origins of life, meaning, and the universe itself* (New York: Dutton, 2016).

5. Sean Carroll, *Consciousness and the laws of physics* [preprint 2021, 5]. http://philsci-archive.pitt.edu/19311/.

A perspectival shift is all that is needed to get us out of the metaphysical hole we seem to find ourselves in.

But I cannot escape feeling that some sleight of hand is part of this line of argument. It feels as if some (presumably unwitting) misdirection is going on—as if the primary problem has been circumvented or even denied, rather than confronted. We start on the terrain of particle physics but shift to arguments at the level of human psychology, all aimed not at whether organisms can choose their actions but at the different question of whether we can ascribe moral responsibility.

In addition, these lines of argument use some concepts that simply may not apply in a deterministic universe. How can an agent be the real causal source of an action when everything that happens in the universe is fully determined by what happens at the lowest levels? It doesn't seem to make sense even to talk about an agent doing A in some situation, based on its own reasons, with the notion that it would have done B if the situation had been different. It is fine to say that it would do B in a different situation, but not that it would have done B if the situation had been different because in a deterministic universe, it never could have been different. Such "counterfactual" scenarios just don't arise.

Indeed, how can we justify using the word "doing" in this scenario at all? Is an organism *doing something* if the events that comprise the supposed action are actually just physical processes inevitably happening within it? "Doing something" actively seems to require the real possibility of doing something else. And how would reasons enter the picture? In a deterministic universe, things just happen, then other things happen, then other things happen, and so on. Nothing *does* anything—certainly not for a reason.

It's not obvious that there are even any causes in a fully deterministic universe. There are forces at work, of course, and there is a sequence of events through time, but it is debatable whether the common conception of causation applies at all. For one thing, the fundamental laws of physics are time reversible: nothing in them specifies that events should flow from past to future. It's difficult to maintain a coherent notion of causation if effects can precede their causes. But more generally, there seems no way to connect specific effects with specific causes in such a globally deterministic scenario. In ordinary parlance, when we identify something (X) as a cause of something else (Y), what we mean is that if X had not been the case, then Y would not have happened, assuming everything else in the scenario was unchanged. A cause in this way of thinking is *a difference that makes a difference*. But, again, this kind of counterfactual thinking simply does not apply in a deterministic universe: nothing could have been different, nor could individual causes be isolated from the general progression of events. Everything just happens.

More fundamentally, the thought experiments on which compatibilism is founded *assume the existence of agents* and then make various arguments for how we can hold them responsible for their behavior in a deterministic universe. But it's not at all clear how agents could ever be part of such a universe. Why and how would they evolve when evolution requires adaptation and selection, processes that themselves seem to require some random variation and the possibility of causal influences inhering at higher levels? If the lowest-level interactions are causally comprehensive, why would there exist configurations of matter that seem to exist precisely because that configuration embodies actionable information in a way that *causally influences action* and provides a selective advantage?

Ultimately, these kinds of compatibilist arguments build an "as-if" picture of human free will that is deemed sufficient for justifying moral responsibility while side-stepping the more fundamental problem that agency—the power of an organism to choose an action at all—is ruled out by physical predeterminism. Fortunately, those arguments are largely moot, because this kind of determinism is refuted by quantum physics itself.

## The Swerve

In his theory of atoms, Democritus emphasized a picture of necessity. As he phrased it, atoms always "fall" in "straight lines"; that is, all of their states, from past through present to future, are already determined and they never stray from that predestined path. It was already clear, even

2,300 years ago, the consequences that such predeterminism would have for the ideas of free will and moral responsibility. In response to these difficulties, Epicurus—a contemporary of Democritus—argued that it could not be the case that atoms always follow these predestined paths. *Taking the choice of action as an observable fact,* he argued backwards that atoms occasionally must randomly "swerve" from such paths; otherwise, the future would already be written, and no real choice could exist.

More than two millennia later, this swerve was confirmed by discoveries in quantum physics. As discussed earlier, elementary particles, such as electrons or photons, can be thought of as the localized manifestations of quantum fields—as their "avatars," in the words of physicist Frank Wilczek.[6] These particles can vary in properties, like their momentum and the energy they carry, but not over a continuous range of values (like height); instead, they can take one of a set of possible, discrete states (like shoe size). For example, the electrons orbiting the nucleus of an atom can be in various discrete energy states but cannot take values in between them: instead, they can jump from state to state.

In 1925, the young physicist Werner Heisenberg figured out a way to handle the mathematics of quantum systems when you don't necessarily know which state each particle is in. The trick is to treat them, mathematically, as if they are in every state at once by performing calculations over a big matrix of possible values (more precisely, the observable values of the intensity and wavelength of light emitted if the electron moved from any one of these states to any of the others). In his 2021 book *Helgoland,* physicist Carlo Rovelli beautifully describes how Heisenberg came to this idea and was able to predict experimentally observable parameters with amazing precision.

Shortly thereafter, Erwin Schrödinger—whom we met in his later years in Dublin in chapter two—devised a completely different way to treat quantum systems that turned out to be mathematically equivalent to Heisenberg's method. Schrödinger treated the particles as if they

6. Frank Wilczek, *Fundamentals: Ten keys to reality* (London: Penguin, 2021), 386.

were waves. The physics of waves was well understood, and there were various reasons to think that quantum particles had some wavelike properties. He came up with a mathematical *wave function* that could describe and predict how a quantum system would evolve through time, from state to state; it also accurately reproduced experimental findings. The trouble is that when we actually see electrons, they do not seem like waves at all: they really do behave like discrete particles. It's as if they only behave like waves when we're not looking at them!

It was left to another pioneer of quantum mechanics, Max Born, to provide an interpretation of these mathematical abstractions. He reasoned that what both Heisenberg's matrices and Schrödinger's wave function were representing was *a map of probabilities*; specifically, the probability of finding in your experimental setup an electron with certain properties at a certain position in space when you interact with it. These probabilities turn out to match with amazing precision the actual outcomes of experiments.

You can take all the relevant values at some initial time, $t$, and plug them into this equation and calculate precisely the values at some later time. Those values evolve according to the wave function in what appears to be a completely deterministic fashion. Some people take this to mean that quantum physics is totally deterministic. *But what those values represent are probabilities.* As soon as you do an experiment to test these predictions, what you find is that the values "collapse" to one of the possible states. If you do lots of experiments, you discover that the statistics perfectly match the calculated probabilities. But in any single experiment, the singular outcome is utterly unpredictable. It does not seem to be predetermined by anything that we know of. The Schrödinger equation describes the behavior of quantum systems perfectly when they are acting like waves, but they stop acting as waves as soon as the particles interact with something. Which outcome we see at that point seems to reflect the random playing out of the underlying probabilities.

Is this genuine randomness at work, or does it simply reflect our ignorance? A popular theory—or family of theories—suggests that maybe we are missing some important physical variables that actually determine the outcome for each individual instance. A lot of work has gone into looking for such "hidden variables," but their existence seems to have been definitively ruled out by experiment, based on theoretical work by the Belfast-born physicist John Bell and subsequent experimental tests of his theory. If they exist, they are so well hidden from the rest of reality that they can never have any influence on anything.

An alternative theory is the *many worlds hypothesis*, first proposed by Hugh Everett in 1957. This is frankly much wackier yet oddly popular. It takes the Schrödinger equation completely at face value and assumes that *all the probable outcomes* predicted actually happen—but just not all in our universe. Every time some cloud of probabilities takes on a single definite value, the same cloud takes on one of the other possible values in a new universe created by that event. If you observe an electron at position x, then, at the same time, *a new you is created in a new universe* observing the electron at position y. You can imagine, with all these subatomic particles interacting with things all the time, that the number of parallel worlds rapidly goes to infinity.

In this model, there is supposedly no randomness: everything proceeds completely deterministically, because *everything that is possible* proceeds. The problem is that this model does not explain which universe you will happen to find yourself in. The argument is that a version of you is in all the branching universes, but there is nothing to determine which universe *you*—the version reading this book right now—actually experiences. So, randomness is not eliminated by this fantastical contrivance of an ever-expanding multiverse—it is just pushed out of sight.

We are left with what is the prevailing view in physics: that the indeterminacy observed in the evolution of quantum systems is real and fundamental. It does not just reflect our lack of knowledge—this is how the universe behaves, at least at these very small quantum scales. Epicurus was correct: atoms (or at least the subatomic particles) really do swerve. But what does this mean for big things like us? Does all this randomness at the quantum level actually make any difference at the so-called classical level of Newtonian mechanics?

## Uncertain, the Future Is

Newton's laws of motion seemed to be completely deterministic—so much so that Laplace could imagine his demon visualizing the entire history of the universe: past, present, and future all laid out at once in a single timeline. Yet quantum mechanics seems to have some fundamental indeterminacy right at the heart of it. How can these apparently contradictory views be squared? After all, classical systems are still made up of subatomic particles. What marks the transition between things that behave like quantum systems and those that obey classical laws?

The traditional view in physics is that when systems get large enough, when they have enough components interacting together, the quantum properties of individual particles dissipate across the whole system in such a way that their probabilistic nature gets averaged out. This process is known as *quantum decoherence* and is thought to reflect the progressive "entanglement" of any given particle with all the other components of the system in a way that forces the particle to adopt a singular definite state. It is often argued that this process means that any traces of quantum indeterminacy get averaged out when some ill-defined threshold is crossed, from the microscopic world of quantum weirdness to the macroscopic terra firma of classical physics.

In this view, quantum indeterminacy would not have much relevance for our discussion on the possibility of agency and free will, which hinges on whether much larger systems are deterministic in nature. For some systems, this is likely true. The systems where Newtonian mechanics works well to predict outcomes are those like the orbits of planets—ones where only a few components are interacting and the dynamics are *linear*. This means that a small change in the value of some parameter at one time point will lead to a correspondingly small change in the predicted value at a later time point. At the scale of classical measurements, such systems really are deterministic well into the future.

However, most systems are probably not like that. If we take the weather, for example, we see exactly the opposite property. The atmospheric system is quintessentially *chaotic*. This means it has stark and unpredictable *nonlinearities* in its dynamics—small changes in some parameters can lead to enormous changes in how the system evolves through time. Any system that has lots of components all interacting at once, with potentially amplifying feedback loops, shows this kind of behavior. Living systems, of course, fit this description perfectly.

But many physical systems do as well. Indeed, the observed organization of matter in the universe depended on the presence of quantum fluctuations in a phase of "cosmic inflation" that preceded the "hot" Big Bang. As the universe was rapidly expanding, it was initially a homogeneous field of energy. Without random quantum fluctuations that broke up this symmetrical field, matter and energy would have been so evenly distributed that pretty much nothing would have happened. Instead, these tiny blips introduced enough inhomogeneity into gravitational fields that galaxies and stars and planets could form. The importance of some randomness in the early origins of the universe itself was also recognized with impressive prescience by Epicurus.

That is an example of quantum randomness acting as a determining factor in how a macroscopic system (the whole universe) actually evolved through time or, more broadly, allowing that evolution to occur at all. But the more important point is that indeterminacy at the lowest levels can indeed introduce indeterminacy at higher levels. This fits with other views on what the quantum-to-classical transition really reflects and on the supposed determinacy of classical systems (see Figure 7.1).

Physicists Lee Smolin and Clelia Verde have proposed that the quantum-to-classical transition does not reflect spatial scale at all but rather the flow of time. In fact, they argue that what we experience as the present is simply the period in which the indefinite becomes definite. In this view, all systems have quantum properties in the future. That is, the properties of the individual particles are probabilistic—they are inherently undefined. It is only when the particles interact that those properties resolve into definite values. What we call "the present" is that period of transition from a future that is indefinite, in which multiple possibilities exist, to a past that can no longer be changed. This process is not instantaneous: it takes time. The present, therefore, has some duration. Rather than eliminating quantum indeterminacy at classical

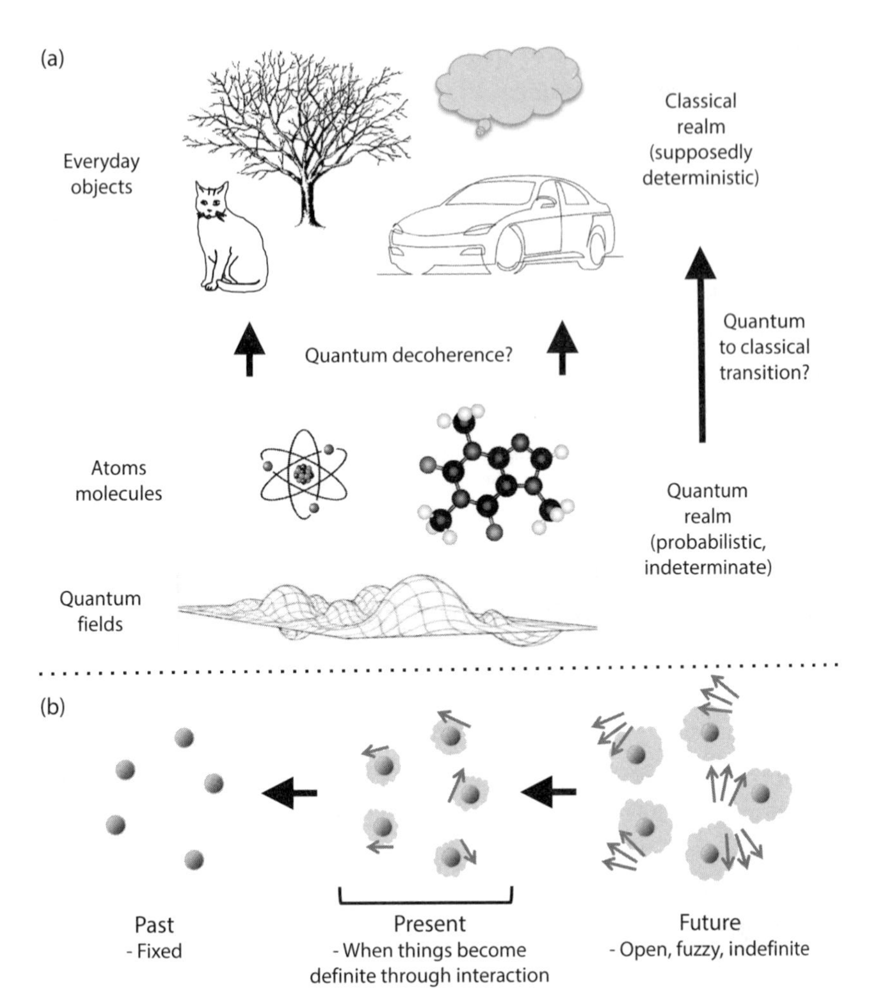


FIGURE 7.1. Quantum to classical transition. **(a)** A prominent interpretation of physics admits indeterminacy at the smallest scales of matter—quantum fields, particles, and atoms—but argues that this indeterminacy gets so spread out through all the interactions of these particles in large objects that systems at the classical level (above some arbitrary scale) behave perfectly deterministically. **(b)** An alternative interpretation views this relationship temporally, rather than spatially, arguing that indeterminacy is a fundamental feature of both quantum and classical parameters *in the future*. This indeterminacy is resolved through interactions between fields, particles, or larger objects, with such interactions defining the period we experience as the present. The past is then fixed, while the future remains open, fuzzy, and undetermined.

levels, this process constantly *introduces it* through the random realization of possibilities. A parallel view, articulated by Nicolas Gisin and Flavio del Santo, claims that this future indefiniteness is not restricted to quantum systems but applies to physical parameters at larger scales as well. They argue that the apparent determinacy of Newtonian mechanics rests on a crucial assumption: that the numerical values of the relevant physical parameters are given with infinite precision, all at once. But that assumption of infinite precision comes up against a hard limit—the amount of information that can be physically encoded in any finite amount of physical space. Under strict determinism, the information about all the particles of the universe right now, at this moment, would somehow have to have been present at the moment of the Big Bang. And the same would be true for every other moment of time, requiring an impossibly infinite amount of information.

Gisin and del Santo argue that, beyond some decimal point, the numbers describing the physical parameters of classical systems in the future simply become indefinite. It's not just that we don't know what they are—the universe also doesn't know what they are, not yet. The numbers are not given with infinite precision all at once; instead, they *evolve as processes through time*. Again, it is through interaction that these parameters take on a definite value, paralleling the "collapse" of quantum possibilities. It's not that these parameters can take any value; they're not completely random. They are constrained, often tightly, by the physics of the system in question—they are, however, *not completely constrained*.

The upshot of these views is that the future is open: indeed, that is what makes it the future. Because we only inhabit the present we don't experience this indeterminacy firsthand. We can't directly experience the future as it currently is (if it even makes sense to think in those terms); we can only predict what it will be like when we get there (or when it gets here). By definition, at that point it will be the present, and things will have become definite, in a kind of just-in-time reality. If we could really glimpse the future, we would see a world out of focus. Not separate paths already neatly laid out, waiting to be chosen—just a fuzzy, jittery picture that gets fuzzier and jitterier the further into the future you look.

These models of the transition from indefinite to definite are controversial but have some intuitive appeal in that they remove the arbitrariness that otherwise creeps in when thinking about the quantum-to-classical transition. Moreover, they contradict the idea of randomness as some extra, positive factor that somehow enters reality, like a miracle, and pokes things around—an uncaused cause that seems to affront the scientific sensibility. Instead, these models argue that physical indeterminacy, *as a negative*—the lack of complete precision of physical parameters at any instant of time—is a default state of physical objects, increasing the further away you go from the present. It's not correct to think of some events as determined and others as random: all events proceed from indefinite to definite. That may be, in fact, what defines events.

This harkens back to another contribution from Werner Heisenberg—his famous "uncertainty principle." This states that you cannot measure both the position and the momentum of a subatomic particle at the same time with perfect precision. In fact, the more precisely you measure the position, the less precise will be your estimate of its momentum, and vice versa. (The same relationship holds for a number of other "conjugate" pairs of physical parameters, like the energy of a photon and the precise time it was released from an atom.) The name of this principle is a bit unfortunate, because it suggests it merely has to do with the knowledge that an observer can have. In fact, it reflects a much more fundamental truth. The wavelike nature of quantum particles means that their position and momentum are both probabilistic and are mathematically related in such a way that narrowing the range of probabilities for one of them necessarily increases it for the other. In fact, this is true even for classical-sized objects: it's just that the margin of error—the fuzziness of these parameters—is negligible relative to the size of the object. So the limit to what an observer can know reflects the indeterminacy of the system itself. Indeed, Heisenberg's first name for this was the "indeterminacy principle."

As Master Yoda put it, "Difficult to see. Always in motion is the future." Physicist Frank Wilczek states, along similar lines, "According to the principles of quantum mechanics, anything that can move, does move, *spontaneously*" (my emphasis).[7] Under the models presented earlier, this indeterminacy gets resolved as particles catch each other mid-jitter and interact to form some new state, the components of which start jittering all over again.

If all of this discussion makes your head spin, you're not alone. Richard Feynman famously said, "If you think you understand quantum mechanics then you don't understand quantum mechanics." There is an entire edifice of mathematics and theory that agrees extraordinarily well with experimental observations; we just don't understand what it means for the nature of reality. But one thing is clear: the current complete state of a physical system—all the relevant properties of all the particles, specified with as much precision as is physically possible, plus the low-level laws of physics—*does not* fully predict the next state of the system. That model of complete physical predeterminism, with a single timeline from the start of the universe to the furthest point of the future, is dead.

## Causal Slack

So, where does that leave agency and free will? On the face of it, it doesn't really help. If all of physics were really deterministic, then possibilities and choices would not exist, and *you* would not be in control of your actions. (I think you would not exist at all, in fact.) But just adding some randomness does not obviously solve the problem. If my actions are controlled by random physical events at the level of subatomic particles in my brain, then I am no more in charge of them than if they were fully physically predetermined. That argument is perfectly valid, but it misses the wider point.

The idea is not that some decisions are determined (driven by necessity) and others are driven by chance. No, the really crucial point is that the introduction of chance undercuts necessity's monopoly on causation. The low-level physical details and forces are *not causally compre-*

7. Wilczek, *Fundamentals*, 80.

*hensive*: they are not sufficient to determine how a system will evolve from state to state. This opens the door for higher-level features to have some causal influence in determining which way the physical system will evolve. This influence is exerted by establishing contextual constraints: in other words, the way the system is organized can also do some causal work. In the brain, that organization embodies knowledge, beliefs, goals, and motivations—*our reasons for doing things*. This means some things are driven neither by necessity nor by chance; instead, they are *up to us*.

As Epicurus put it, "Necessity, introduced by some as the absolute ruler, does not exist, but some things are accidental, others depend on our arbitrary will. Necessity cannot be persuaded, but chance is unstable. It would be better to follow the myth about the gods than to be a slave to the [destiny] of the physicists."[8]

Epicurus's real target was thus *reductionism*, not merely determinism. His swerve was necessary to give the will some room in which to operate. The contemporary philosopher and mathematician George Ellis similarly argues that physical indeterminacy creates *causal slack* in physical systems, which opens the door for what is known as "top-down causation." Put simply, this is the principle that the way a system behaves depends on the way it is configured, which can constrain the lower-level components and functionally select among patterns of those components. In living organisms, that configuration itself is the outcome of selection over millennia and over the lifetime of the organism itself, on a timescale of seconds to hours to years. This highlights another key principle, which also diverges from a reductive, comprehensively bottom-up view: causation is not wholly instantaneous.

This principle brings us to the second variety of determinism referred to earlier: causal determinism. Simply put, this is the idea that every event is necessitated by prior causes. Things don't just happen, nor can things be the cause of themselves. In some framings, this is essentially identical to physical predeterminism, but only if all the causes are

8. As quoted by Karl Marx, "The Difference Between the Democritean and Epicurean Philosophy of Nature," in *Marx-Engels Collected Works, Volume 1* (Moscow: Progress Publishers: 1902), 21.

deemed to be located at the lowest levels of reality, such that events are solely necessitated by antecedent events and conditions, together with the (low-level) laws of nature. If this is not the case—if higher-order features and principles can exert causal power in a system—then causal determinism poses no threat to concepts of agency and free will. The agent itself can be the cause of something happening.

Just a few decades before the time of Epicurus, Aristotle articulated a theory of causality, which defined four causes or types of explanation for how natural objects or systems (including living organisms) behave. The *material cause* concerns the physical identity of the components of a system: what it is made of. The *efficient cause* concerns the forces outside the object that induce some change. On a more abstract level, the *formal cause* relates to the form or organization of those components. And—finally—the *final cause* refers to the end or intended purpose of the thing. Aristotle saw these as complementary and equally valid perspectives that can be taken to provide explanations of natural phenomena.

Regrettably, in the early 1600s, Francis Bacon, one of the fathers of the modern scientific method, argued that science should be concerned solely with material and efficient causes; that is, with mechanisms. He consigned formal and final causes to metaphysics, or what he called "magic." Those attitudes remain prevalent among scientists today and not without reason—that focus has ensured the phenomenal success of reductionist approaches that study matter and motion and deduce mechanism.

But such approaches are focused exclusively on answering "how" questions. If I want to understand how my pressing the keys on my laptop right now gets translated into words appearing on the screen, I could get a complete description of the mechanisms involved and the sequence of causal events and interactions (Aristotle's material and efficient causes). But that description would not be a complete explanation of what is happening. There is a "why" question to be answered as well. Tapping on the keys only produces the desired result because the computer is *designed and programmed that way*. It has a function (a final cause or purpose) that is built into its structural configuration (which becomes the formal or organizational cause). That process of design and construction happened a long time ago, but it is just as important a cause of the words appearing as my tapping my fingers or the electrons whizzing through the computer's circuits. And, of course, I have a purpose in this activity too—to convey ideas to you, the reader.

Many scientists are suspicious of "why" questions. They smack of cosmic teleology, the idea that things in nature are the way they are because they have been driven that way by some overarching purpose, thereby implying the hand of some intelligent designer. There's no evidence that this is the case; or as Laplace replied to Napoleon, who asked him where God was in his scheme, there is "no need for that hypothesis." However, although the universe itself may not have purpose, living organisms certainly do. That is, in fact, their defining characteristic.

## Better Living by Design

As we saw over the preceding chapters, the processes of natural selection let functional design modifications accumulate. The logic behind this is at once trivial and relentless: systems incorporating dynamics that tend to favor their own persistence will necessarily out-persist ones that do this less well. If you have a population of such systems (living organisms, in this case, although the logic works just as well for computer programs or other entities), then you simply add variation, apply selection for persistence (or whatever functionality you want), and repeat. And repeat, and repeat. The outcome is that those self-sustaining dynamics become embodied into the structure of the system. Functions evolve, because subsystems work in ways that ultimately increase survival and reproduction.

It's important to emphasize that the resultant functionalities—the specific architectures that carry out various operations—do not just *emerge from* the lower-level components, any more than the strategies of a football team emerge from what the individual players are doing. The relationship is precisely the converse: the functional architectures or the strategies have an independent origin and *constrain* the actions of the individual components. Paradoxically, as philosopher Alicia Juarrero argues, constraining the actions of components allows the whole system greater freedom to pursue higher-order goals. We pay football coaches so much because organization matters.

Functional architectures do not have to be invented by evolution—they are discovered. There are abstract principles of mathematics and engineering that just "exist" or hold true, independent of any physical substrate. These principles do not emerge from the bottom up. They are every bit as fundamental as the equations describing the core theory of physics, and they are independent of them in that they cannot be derived from those equations. If you want to build a filter or an oscillator or a coincidence detector or an amplifier, there are certain functional designs that work, regardless of whether the system is embodied in electrics or electronics or networks of genes or neurons. Evolution is effectively a search for these kinds of functional architectures—for any way of configuring the components of an organism that helps it stay alive.

The algorithm of natural selection—the simple steps of reproduction with variation and selection, repeated thousands and thousands of times—acts like a ratchet. Each new advance is held in place, while new variants are combed through for ones that add another small improvement. In this way, the randomness that generates the new architectures that are tested is translated into a directional, progressive process. (Note that the algorithm of natural selection itself is a good example of a functional principle that simply exists, independent of the low-level laws of physics.) The key property of systems undergoing natural selection is that they incorporate information from their history into their own physical structures. This ability allows them in turn to act on information in the present moment. Causation is thus extended over time.

Consider a bacterium that encodes a receptor protein that can bind specifically to some molecule in the surrounding environment, and couple that binding to a directional movement. You could say that at any given moment, the binding of the molecule causes the bacterium to move a certain direction. And you could trace all of the physical interactions of atoms and molecules that mediate this chain of events: they make up the mechanism at play. But the binding of the molecule is only what philosopher Fred Dretske calls the *triggering cause*. To understand the system, we must also consider the *structuring cause*—what it is that led the system to be structured in such a way that when it detects that specific molecule, it moves in a particular direction.

That structuring obviously arose over a very long timescale through all the iterative feedback mechanisms of natural selection. But it is every bit as much the cause of what's happening as the momentary biochemical interactions. The triggering cause answers the "how" question; the structuring causes answer the "why" question. Of course, as anyone who's had young children knows, there is always another "why" question lurking behind the first one. In this case, the reason why the bacterium has that structure is that it is adaptive, or to be more precise, it has proven to be adaptive in the past. As such, it's a bet that natural selection makes: the future will look like the past, and the information represented by this particular molecule (say, that a food source is nearby) will continue to be the case and continue to be useful.

Agents can act with causal power in the world because, biologically speaking, they have been paying attention. This is not a free lunch. Like potential energy, living systems act as stores or capacitors of potential causality. The difference is that the content to be stored is information; specifically, causally effective information. I described earlier the idea that a cause is "a difference that makes a difference"—a concept that only makes sense in a nondeterministic universe. I borrowed that phrase from Gregory Bateson, who actually used it to define information. Causal influences acting on the organism leave a physical trace: a record of causal relations in the world internalized in the structures of the organism itself. The resultant configurations can then be used to do causal work by setting criteria for action depending on current information. As we will see in the next chapter, the details at the lowest levels cease to matter: what matters is the overall pattern and what that pattern means. Information thus has causal power in the system and gives the agent causal power in the world.

In summary, there is nothing in the laws of physics that rules out the possibility of agency or free will, a priori. The universe is not deterministic, and as a consequence, the low-level laws of physics do not exhaustively encompass all types of causation. The laws themselves are not violated, of course—there's nothing in the way living systems work that contravenes them nor any reason to think they need to be modified when atoms or molecules find themselves in a living organism. It's just that they are not sufficient either to determine or explain the behavior of the system. In the next chapter, we will look at how this indeterminacy manifests in neural systems and how decision-making systems evolved to harness it.

# 8

# Harnessing Indeterminacy

We saw in the last chapter that the state of a physical system at any given point in time, plus the elementary laws of physics as captured by the equations of the core theory, *do not* in fact predict the precise state of the system at some subsequent time point. There is substantial indeterminacy at play, which means that the evolution of many types of systems—especially ones with complex or chaotic dynamics—is unpredictable in practice and inherently undefined in principle.

In the brain, this indeterminacy manifests as variability in the cellular processes and electrical activity of neurons. This generates what engineers call *noise* in neural populations: random fluctuations in the very parameters that are used to transmit signals. This noise presents nature with a problem: it is difficult to build structures capable of complex cognitive operations out of individually unreliable components. But organisms also capitalize on this underlying variability; in fact, it is absolutely essential to enable flexible behavior. Crucially, it breaks what Epicurus called "the treaties of fate," under which the behavior of the organism would simply reflect the inevitable transitions from one physical state to the next. Instead, the brain has evolved to take advantage of the noisiness of its components to allow the organism to make some decisions itself.

## It's Noisy in There

In previous chapters I described neurons as marvels of nature's design, exquisitely crafted to collect and process and distribute information. But from an engineering perspective, they present some challenges. They are made of wet, jiggly, incomprehensibly tiny components that jitter about constantly, diffusing around at random, bumping into each other, engaging in transient molecular interactions, shifting their conformations, and continually being chemically modified, transformed, broken down, and remanufactured. How can you make a reliable information processor out of this kind of messy wetware?

To quickly review, when a neuron fires a nerve impulse—an action potential or spike—it travels as an electrical wave down the axon, the output fiber of the neuron, to the synapses, its connections with other neurons. The synaptic terminals at the ends of the axon are highly specialized structures that are filled with little vesicles, bubbles of membrane prepackaged with neurotransmitter molecules that are ready to be released to communicate to other neurons. At each such synaptic terminal, the arrival of the spike triggers the opening of calcium channels in the membrane, allowing calcium ions to rush into the synapse. These calcium ions are bound by proteins that sit in the synaptic vesicles, and when the local concentration of calcium gets high enough, it triggers the fusion of the vesicle with the cell membrane.

The bubble then pops, releasing its contents into the narrow cleft between the sending and the receiving neurons. The neurotransmitter molecules can then be detected, before they diffuse away, by specialized receptor proteins on the receiving cell's membrane that open ion channels, letting sodium ions flow into the neuron at that point. If enough ions flow in, this can trigger a new spike in the receiving neuron. In the meantime, the neurotransmitter is recycled back into the sending neuron's synaptic terminals, along with the extra membrane from fusion of the vesicles, while all kinds of transporter proteins pump the various ions back out of the cell to restore the electrical potentials required to drive the process.

Now, if you were setting out to design a system like this, you might want to ensure that when a neuron fires a spike, it necessarily results in the release of neurotransmitter at its synaptic terminals. Otherwise, it's a damp squib: whatever information was represented in the spike itself is lost if it's not communicated to downstream neurons. Similarly, if neurotransmitter is released but not reliably detected or transformed into electrical activity by the downstream neurons, then again, that information effectively evaporates. The problem is that all the protein components in these processes are subject to noise and random "thermal" fluctuations. The binding of proteins to each other or to chemical ions depends on their relative concentrations, but in a probabilistic sense. It is an equilibrium process, with binding and unbinding both happening over any period of time—higher concentrations just make the bound state more likely. The same is true for the conformational changes that proteins undergo: they are also probabilistic. Indeed, at these small scales, these events are even subject to quantum indeterminacy. As a result, ion channels sometimes fail to open even when neurotransmitter is bound or the electrical potential is high, and synaptic vesicles sometimes fail to fuse even when the concentration of calcium ions in the synaptic terminal is high. Conversely, ion channels open spontaneously and synaptic vesicles fuse spontaneously at a certain frequency. All these factors can, at least in theory, degrade the reliability of signal transmission within each neuron and from one to the next.

Of course, nature has solved this problem—otherwise I wouldn't be able to write this sentence and you wouldn't be able to read it. Noise in individual molecular components can be buffered by simply adding more of them to make the neurons more sensitive to signals and less susceptible to independent, random fluctuations in specific proteins. The problem with that approach, from the organism's point of view, is that it's really expensive. It takes energy to make those proteins and get them in the right places, and it takes energy to keep the balance of ions inside and outside neurons at the appropriate levels. Moreover, these proteins take up space in the membrane: adding more and more of them can only be accomplished by making the neuron bigger and increasing the diameter of its axon and dendrites. This causes problems for the electrical conductance of the membrane, it increases the cost of sending spikes, and it also means that you can fit far fewer neurons into the same-sized brain. So, there are some practical limits to that brute-force approach to ensuring reliability.

The other method relies on the interlocking dynamics of all the feedback and feedforward loops involved in electrical signaling in the neuron, which collectively can dampen molecular noise and achieve robust signaling. Indeed, some studies have found that, given a certain input, individual neurons very reliably produce a specific series of spikes as outputs. In these cases, electrical current is directly injected into the cell body of the neuron, thus bypassing all the synapses and dendrites that normally collect inputs. Under more natural conditions, recording from neurons in an intact animal typically reveals considerable variability from trial to trial (say of presentation of a specific stimulus). There is a lively debate about the source of this variability; some researchers argue that it is caused by inherent random fluctuations in ion channels and synaptic machinery. Others argue that the variability is not noise at all but rather reflects contextual signals from all the other neurons that the recorded neuron is attached to that are driven by what the animal is doing at any given moment.

It seems likely that both claims can be true and may apply to different extents in different neurons and circuits or even in the same neurons under different conditions. Some neurons may be tuned in such a way that a strong signal is required to activate them, thus producing a reliable input–output relationship. Others may be poised much closer to a threshold, where small fluctuations within them can have an effect on their activity. And all of them are talking to each other in ways that can, at the population level, be designed to robustly average out noise or to take advantage of it. Indeed, many neural networks are configured in what is known as a *critical* state, able to adjust their patterns of activity flexibly and rapidly in such a way that small changes can sometimes be amplified to produce quite divergent outcomes.

This is a highly adaptive arrangement. After all, the function of the nervous system is not merely to reliably transmit signals from one neuron to the next—that would be pointless, when you think about it. And it's not to precisely carry out logical operations or predefined algorithms to output a "correct" answer, like a digital computer. The function is to help an organism adapt to ever-changing circumstances in its environment. If the parameters of the problems facing the organism were fixed and fully known ahead of time, this challenge would be much easier; it could be met by highly specified circuits carrying out highly specified tasks. But the world is not like that—the challenges facing organisms vary from moment to moment—and the nervous system has to cope with that volatility: that is precisely what it is specialized to do. A good strategy in such a world is for an animal to vary its own behavior sometimes. Even in a situation where things are going well, it pays to keep checking that the world still is operating the way you think it is by occasionally engaging in exploratory behaviors to update the model of both the self and the world with current information.

This capacity is important in even the simplest choices of the simplest organisms. We saw in earlier chapters how bacteria can navigate away from danger and toward food sources by switching between linear motion in a particular direction and making a random tumble. If the bacteria are onto a good thing—if their movement is taking them up a concentration gradient of a food molecule, for example—then they spend more time in the linear motion mode and less time tumbling. If they're not getting such positive signals, they initiate more random tumbles, followed by forays into random directions.

It turns out that the tendency to initiate random tumbles *is itself subject to some randomness*, varying even among genetically identical bacteria. This is influenced by how much chemosensory proteins the bacteria make; this process is susceptible to molecular noise within the bacterial cell. Differences in the DNA sequence of the genes encoding these proteins affect how noisy the process is. It seems that bacteria are selected to have a certain degree of noise in this system, perhaps as a way of hedging their bets. Even across a population of genetically identical bacteria, some are making more of those proteins at any given moment and are therefore more likely to continue in "exploit" mode, whereas others may be more likely to go into "explore" mode. Natural selection has thus calibrated the level of noise in the system to help simple organisms adapt flexibly to a changeable environment. The same is true in multicellular organisms, where noise in the nervous system loosens stimulus–response mappings and enables behavioral flexibility.

The other crucial function of nervous systems, of course, is to allow organisms to learn, to reconfigure their circuitry to reflect past experience and better anticipate future circumstances. If the neurons were already maxed out—if the connection weights were all set to 100 percent strength—then no learning would be possible. By contrast, neurons operating in a responsive range of probability of signal transmission can have that probability modified up or down. Indeed, organisms go to quite a bit of trouble to keep their neurons in the responsive range, renormalizing all synapses during sleep to ensure that a busy day of learning doesn't overfix the system.

From this point of view, we can see that the apparent unreliability of neural transmission at the level of (at least some) individual neurons *is a feature in the system, not a bug.* The noisiness of neural components is a crucial factor in enabling an organism to flexibly adapt to its changing environment—both on the fly and over time. Moreover, organisms have developed numerous mechanisms to directly harness the underlying randomness in neural activity. It can be drawn on to resolve an impasse in decision making, to increase exploratory behavior, or to allow novel ideas to be considered when planning the next action. These phenomena illustrate the reality of noisy processes in the nervous system and highlight a surprising but very important fact: organisms can sometimes *choose* to do something random.

## Being Unpredictable

There are a few good reasons why an animal might want to do something random. First, there may be circumstances—perhaps not uncommon ones—where acting randomly *and thus unpredictably* is directly adaptive. This is a good idea if something is trying to eat you, for example, but readers who play poker may also recognize the benefits of making some decisions at random so that opponents cannot pick up on patterns in your play. Indeed, this was a central realization in the development of game theory by John von Neumann, Oskar Morgenstern, and, later, John Nash: if you are playing opponents who can benefit from predicting your moves, it pays to be random. And life is a game where lots of opponents are trying to predict an animal's moves.

Intrinsic variability is observed in escape behaviors in many animals. For example, cockroaches, when threatened, scuttle off rapidly in what appears to be a random direction. In fact, close analysis shows that the animals select from a restricted and predefined set of possible escape angles relative to the location of the threat. Yet, under the most carefully controlled experimental conditions, the particular angle chosen remains entirely unpredictable. Although we do not know the precise neural bases for this behavior, this phenomenon is consistent with a model of choice that we have seen before, in which distinct action plans compete for control of an output command neuron or region. Under those conditions, even small momentary fluctuations in the neurons comprising the circuit can be amplified by winner-takes-all dynamics, in which each pattern reinforces itself and suppresses the others. In the face of predators trying to anticipate your movements, allowing some randomness in the neural circuits to settle the outcome can be highly adaptive.

Now, you might argue that this randomness is only apparent, reflecting ignorance on the part of the experimenters and an inability to really control all the relevant environmental parameters of a cockroach in a lab setting. As the Harvard Law of Animal Behavior states, "Under carefully controlled experimental circumstances, an animal will behave as it damned well pleases." Who knows what state an individual cockroach is really in at a particular moment when you choose to startle it? The variability in responses may illustrate the integrative, holistic nature of neural control of even simple, supposedly hardwired reflexes, but perhaps the appeal to randomness per se, as a deciding factor, is not compelling.

A more direct demonstration comes from studies of the nervous system of leeches (see Figure 8.1). These simple creatures respond to mechanical touch by initiating one of two modes of locomotion to escape: swimming or crawling. It is possible to isolate their nervous system in a dish and apply precise electrical stimulation that mimics the signals triggered by mechanical touch. When the activity of the population of neurons controlling locomotion is monitored in response to such stimulation, it is observed to settle into one of two competing modes, corresponding to swimming or crawling movements. Despite the tightest control of the conditions of the isolated

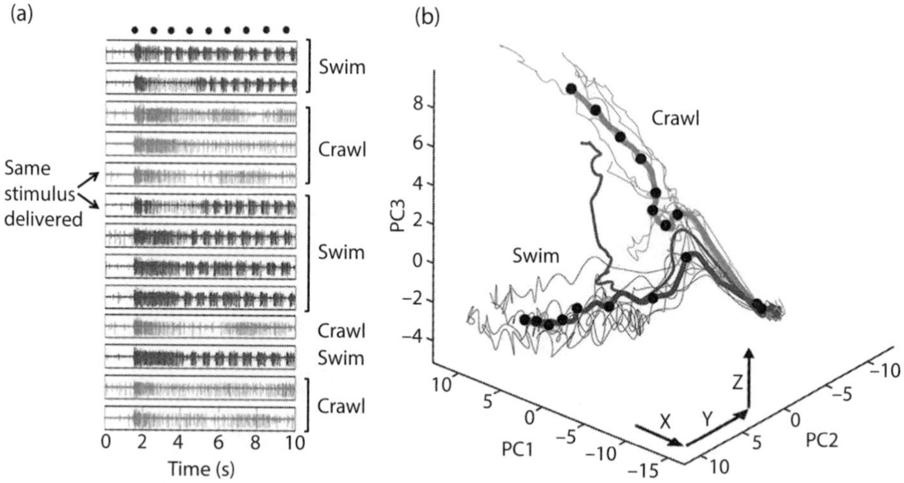

FIGURE 8.1. Neural population firing in leech. (**a**) Each trace shows the activity of a population of neurons in the isolated leech nervous system in response to the identical stimulus. On any trial, one of two alternative patterns emerges unpredictably—one corresponding (in the intact animal) to a swimming movement and the other to crawling. (**b**) By statistically identifying the "principal components" of population states (PC1, 2, 3) it is possible to graphically represent these different trajectories of neuronal population activity in low-dimensional state space.

nervous system and the most precise application of electrical stimulation, the output still seems to vary at random between these two options.

The dynamics at play here are probably a widespread feature of populations of neurons engaged in all kinds of tasks. It's as if each local population of neurons is trying to decide on something—to settle into some state representing a unique solution to whatever it is that population is interested in. Motor neurons, like those in the leech, may be trying to decide on an action. Perceptual neurons may be trying to draw an inference of what is out in the world from ambiguous stimuli, which again may involve a resolution of competing patterns. We can experience that dynamic ourselves, when looking at visual stimuli like the Necker cube (shown in Figure 5.5), where our perception of it oscillates between competing perspectives, seemingly at random.

To understand this process, we need to take a bird's-eye view of the dynamics of whole populations of neurons. Say we have a network with two neurons, A and B. At any time point, let's say for simplicity that each neuron is either active or not active ("on" or "off"). The network can then be in four possible states: both neurons on, both off, A on and B off, or A off and B on. We could plot this on a two-dimensional graph with the activity of A on one axis and the activity of B on the other, providing a handy way to visualize the state of the whole network—basically with four boxes. Now, let's add a wrinkle, by introducing some connections between the neurons—let's say that when A is on, it activates B, and vice versa. Then the states that the whole network can be in will be reduced—only both A and B on or both off will be stable, because if either A or B is activated, both will get activated. In our plot of network *state space,* now only the boxes representing both on or both off will ever be realized; we can say that the network, in its dynamic fluctuations, can "visit" those states, but no others. You can imagine that if we instead have reciprocal inhibitory connections between A and B, that the state when both are on simultaneously would be unstable, while the other three states could be visited (see Figure 8.2).

Now, if we add a third neuron C to this network, we end up with a three-dimensional graph and the states that the network can visit are set, in a more complicated way, by the excitatory and inhibitory connections between all three pairs of neurons. Which states it actually adopts will depend on the inputs to each of the neurons, which means *these states can carry information about those inputs.* But because neurons cannot stay active all the time, the network will also display some emergent dynamics. As one neuron activates, it will excite others with a brief delay; then it will itself switch off with a certain time course as it temporarily exhausts the electrical potential that allows it to fire, thus removing excitatory drive from its targets (or possibly releasing them from inhibition); and so on. In the real nervous system, neurons are not either on or off; instead, their rate of firing can vary continuously, from low to high. Yet the underlying principle holds: the whole network will move from state to state in a way that is continuously informed by its inputs but constrained by its own configuration.

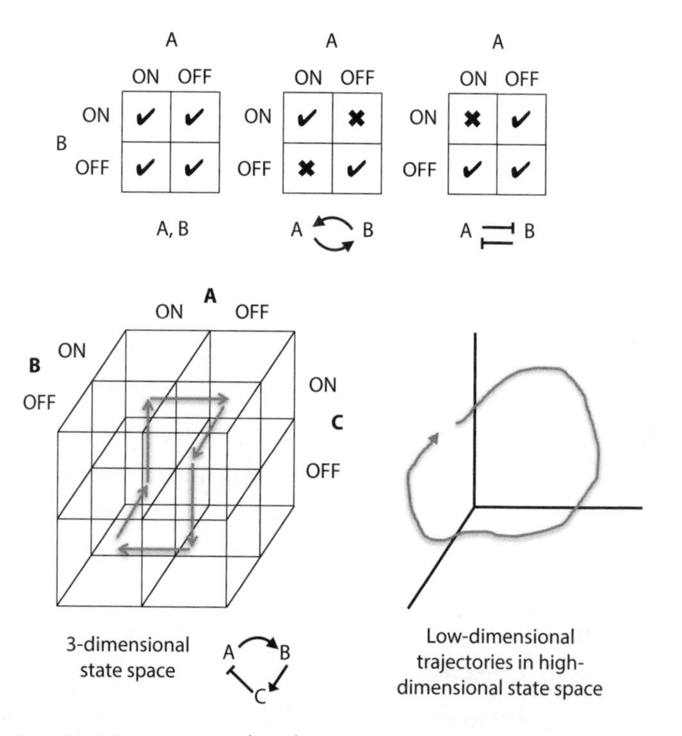


FIGURE 8.2. Population states. **(Top)** The overall state of a system with two components, A and B, which can each either be "ON" or "OFF," can be represented as a grid with four possible states. If A and B are independent of each other (left), all four states may be stable. If A and B mutually activate each other (middle), then only two states are stable (both ON or both OFF). If there is mutual inhibition, then three states are stable (both ON is unstable). **(Bottom)** If a third element, C, is added **(left)**, then the state-space becomes three-dimensional; however, activating or inhibiting relations between the elements will similarly shape which population states are stable, as well as possibly dictating a dynamic trajectory that the population will follow through time. In a system with some arbitrarily large number of elements, *n* **(right)**, there may still be emergent states and trajectories that can be mapped in a much lower-dimensional space.

Now, even though it becomes impossible to visualize, there's nothing to stop us from adding more dimensions to this mathematical network, ending up with as many dimensions as there are neurons (equal to some number, *n*). You can imagine then the exponential explosion of possible states in this "*n*-dimensional" space. But now if we look at the dynamics of activity of the whole network, what we find is that only a small subset of that vast space of possible states is actually visited because of the constraints imposed by the configuration of all the interconnections in the network. The realized patterns of activity in this kind of network can thus be captured mathematically and represented as following trajectories in a very low-dimensional space. Inevitably, *these patterns compete with each other*: because they involve the same components being in different states, they are mutually incompatible.

They are also self-reinforcing, not just in a single moment because of all the reciprocal excitatory connections, but over time. The activation of any synapse does not only lead to a sharp signal in the receiving neuron but also causes a change in its electrical responsiveness that can last for hundreds of milliseconds—a trace that affects the ongoing evolution of its neural dynamics. This means that every time a pattern is activated, it transiently reinforces itself and becomes more likely to be activated again. This feature provides a crucial mechanism that allows neural circuits to *accumulate evidence* over some (still quite fast) period of time. In a situation of perceptual ambiguity, for example, it allows the system to gradually decide what it's seeing, even if the evidence for one particular outcome is not strong enough to drive the network to that state at any given moment. A natural consequence of this dynamic architecture is that, in the absence of driving evidence, the network sometimes accumulates noise and randomly settles into one particular state. And that process can be used by the organism to decide what to do.

## Buridan's Ass

Drawing on a random process may be helpful when an animal is faced with a choice for which the outcome doesn't matter that much or where the options are equally attractive, given the information available. In this situation, the organism has no good reason to choose one option over the other—but it does have a good reason to *do something*, as opposed to being frozen in indecision. All the time it spends vacillating incurs opportunity costs while possibly also exposing it to threats. In many such cases, the right action to take is to look for more information before choosing. But when time is of the essence or when useful information is unlikely to be available, the best thing to do is to just mentally "toss a coin" and go with the outcome.

This is illustrated by a well-known problem in philosophy, named after the fourteenth-century French philosopher Jean Buridan, although the idea dates at least as far back in time as Aristotle. The scenario envisages a hungry donkey placed precisely halfway between two identical piles of hay. Having no reason to choose one over the other and no internal means of breaking the deadlock, the donkey starves to death. We already discussed a similar scenario of the lamprey presented with two attractive stimuli on either side. The optimal behavior in that case is not to average the signals and thus chart a course between them but instead to let the signals compete and ultimately pick one or the other. A perfectly reasonable and highly efficient strategy is to let such decisions be settled by noise—a neural coin toss—rather than wasting time dithering between options of equal value. There's good evidence from recordings in monkeys and rodents that in such situations this is exactly what happens.

This brings us to one of the most widely misinterpreted set of findings in human neuroscience, derived from experiments conducted by Benjamin Libet and his colleagues. They were interested in finding patterns of brain activity that correlated with decision-making processes in humans. An electroencephalograph (EEG) was used to get an indirect measure of brain activity while subjects performed an action. Because neurons in the cortex are radially aligned in parallel with each other, when they are active in a correlated way they generate electrical fields that can be detected by electrodes on the scalp. The inventor of this technique, Hans Berger, thought that these brain waves might be the basis of telepathy. Alas, air is a poor conductor, and the electrical fields don't make it more than a few millimeters from the scalp, but that is just far enough for EEG electrodes to pick them up.

Libet's experimental setup was as follows: participants were instructed simply to lift a finger or flex their wrist whenever they felt like it, explicitly being told to act "on a whim." As they did this, the EEG was recording their brain activity, and they were also watching a clock with

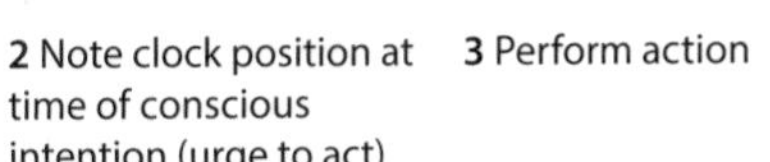
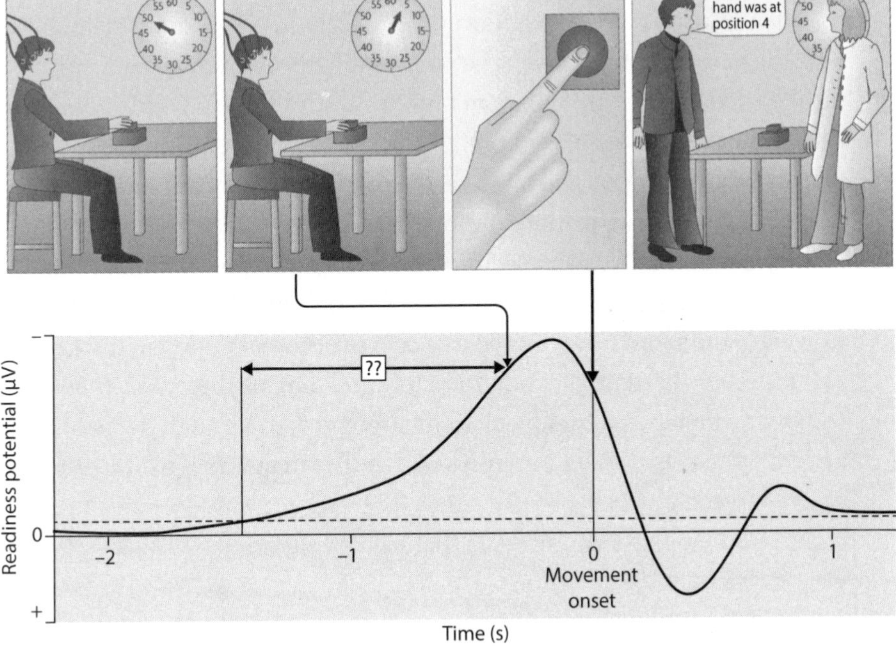


FIGURE 8.3. Libet experiments. In a classic experiment performed by Benjamin Libet and colleagues, subjects were instructed to sit at a table and move their hand whenever they felt the urge to do so. They were also watching a clock and reported when they became conscious of feeling that urge, which typically preceded the actual movement by only tens of milliseconds. However, EEG recordings found that activity detected over the premotor areas of the brain often began to ramp up several hundred milliseconds prior to this consciously felt urge. The investigators inferred that this *readiness potential* reflected the brain "making the decision" to move, rather than the person, with the conscious experience merely reflecting a report of this process having happened.

a second hand. They were instructed to note the time when they first were aware of the intention to move; of course, the actual time of the movement was also recorded (see Figure 8.3). The first findings of the study were completely as expected: patterns of electrical activity (a so-called readiness potential) were detected by electrodes over the brain area responsible for planning movements before the movements were actually done. That makes sense. And the subjects reported the onset of the awareness of "wanting to move" before actually moving, which also makes sense. It was when these timings were compared that the surprise emerged: the brain activity in the motor planning area typically started to ramp up *several hundred milliseconds before* the point at which the participants claimed to be consciously aware of the intention to move. The conclusion was stark: the subjects' brains were deciding when to move and only subsequently reporting it to their conscious minds.

Libet and colleagues concluded, "The brain evidently 'decides' to initiate or, at the least, prepare to initiate the act at a time before there is any reportable subjective awareness that such a decision has taken place. It is concluded that cerebral initiation even of a spontaneous voluntary act, of the kind studied here, can and usually does begin unconsciously."[1] As a description of what's happening at face value in this particular scenario, that conclusion seems fair. However, the researchers also drew wider implications for the question of free will in general: "These considerations would appear to introduce certain constraints on the potential of the individual for exerting conscious initiation and control over his voluntary acts."[2]

Following their lead, the implications of these findings have since been widely extrapolated, way beyond the bounds of the actual experiment, to suggest that we never really make decisions at all, that our brains just do the deciding for us, and that we later make up stories to ourselves to rationalize our actions in some kind of post-hoc narrative. Indeed, these experiments are often cited as conclusive evidence that neuroscience has shown free will to be an illusion. This is, to put it mildly, a drastic overinterpretation.

That is because the design of the experiment makes it effectively irrelevant for the question of free will. The participants made an active and deliberate decision when they agreed to take part in the study and to follow the instructions of the researcher. Those instructions explicitly

1. Benjamin Libet, Curtis A. Gleason, Elwood W. Wright, and Dennis K. Pearl, Time of conscious intention to act in relation to onset of cerebral activity (readiness-potential): The unconscious initiation of a freely voluntary act. *Brain* 106 (Pt. 3) (1983): 640.

2. Libet et al., Time of conscious intention to act, 641.

told them to act on a whim: "to let the urge to act appear on its own at any time without any preplanning or concentration on when to act."[3] They had no reason to want to move their hand more at one point than another because nothing was at stake. And so, it seems they did indeed act on a whim: they (decided to) let subconscious processes in their brains decide, by drawing on inherent random fluctuations in neural activity.

Because EEG data capture lots of ongoing brain activity not related to the movement, it is necessary to average across many traces to see the tiny signal that represents the readiness potential. The way Libet and colleagues did that was to take data for individual "events"—every time the subject moved a finger—and align these traces at that point of movement and then average across them. Looking backward at the averaged, time-locked brain activity preceding the movements, they could then see the readiness potential as a steadily ramping signal. The signal is thus only apparent retrospectively; it is not actually predictive of movement. The interpretation, however, was that the "onset" of this signal—when it first starts to creep up above baseline—*represents the point at which the brain decides to make a movement* and that the subsequent increasing activity represents the process of preparing the movement, culminating in its execution.

However, there is a very different way of looking at these data, developed by Aaron Schurger and colleagues, which is consistent with a totally different interpretation of what is going on: that the organism is making use of ongoing random fluctuations to guide the (utterly inconsequential) choice of when to move (see Figure 8.4). When the EEG recordings are time locked to an externally cued event (a click instructing the person to move as quickly as possible), the signal clearly fluctuates noisily up and down all the time. Sometimes it goes back down again, and the person does not move; other times it happens to reach a threshold, and then a movement is initiated. In Schurger's experiment, if the activity level happened to be nearer the threshold when the person was cued to act, then their reaction times were shorter. These data

3. Libet et al., Time of conscious intention to act, 625.

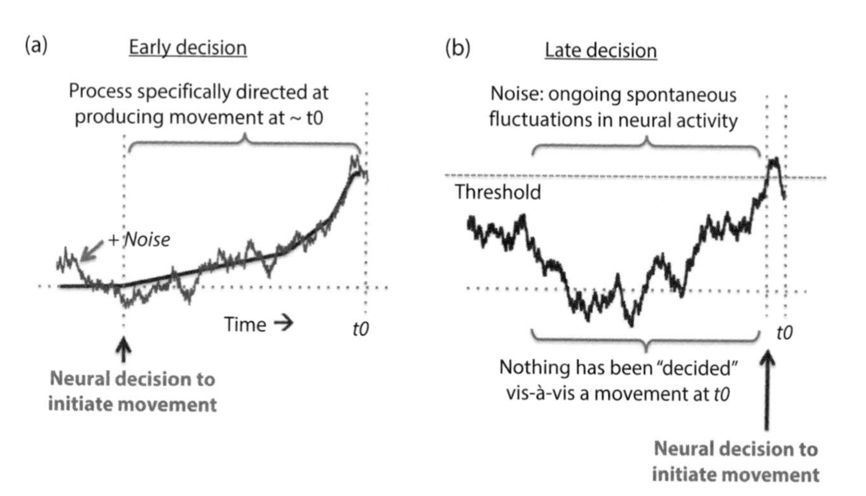


FIGURE 8.4. The Schurger interpretation. Reanalysis of data from Libet-style experiments suggests a very different interpretation. When the EEG recordings are aligned as in the original experiments with respect to the time that movements were made (**a**), the readiness potential becomes apparent. This makes it look like the brain is deciding to move several hundred milliseconds prior to conscious awareness and then the movement inevitably occurs. However, when the EEG recordings are aligned to some arbitrary event (**b**), it becomes apparent that there is constant fluctuation in the signal over premotor areas, which goes up and down through time. It is only when these spontaneous fluctuations pass some threshold that the decision to move is taken, corresponding perfectly with the moment that subjects report being consciously aware of the urge to move (and a window in which they still can exert a conscious veto over the movement).

strongly suggest that the choice uses the same kind of self-reinforcing "evidence accumulation" process for normal decision making, but because nothing is at stake, the subjects allow the process to be driven by random fluctuations. In this interpretation, the commitment to move is made much later, only when the threshold has been reached and exactly within the time window when subjects report becoming consciously aware of such an intention.

This means that the readiness potential is not in fact a signal of the intention to move that occurs long before subjective awareness but rather is an artifact of the way the data are analyzed. A strong prediction of this model is that the readiness potential should not be observed in scenarios where subjects are making a decision that actually matters to them and that they deliberate over. Uri Maoz, Liad Mudrik, and colleagues recently tested that idea in an elegant experiment where subjects had to decide the amount of money to give to two charities. In half the presentations, the decision had no consequence: both charities received $500. But in the other half, the chosen charity received $1,000, and the other got nothing. The charities were selected for each subject from among a set that included those the subjects cared about, giving them an incentive to choose one over another.

The results were very clear: in the trials where the decision was arbitrary, a readiness potential was detected. That is, the movement of the left or right hand was associated with random fluctuations that raised the activity in the premotor cortex above the threshold for initiating a movement. But in the trials where the subjects made a deliberative, consequential choice, no such association was found. Presumably, the subjects were processing information elsewhere in the brain as they were considering the option, and then movement was triggered on that basis.

Overall then, Libet's experiments have very little relevance for the question of free will. They do not relate to deliberative decisions at all, where the readiness potential is not observed. Instead, they confirm, first, that neural activity in the brain is not completely deterministic and, second, that organisms *can choose* to harness the inherent randomness to make arbitrary decisions in a timely fashion. It is likely that we do this all the time, without being aware of it, precisely because that process relates to all kinds of little actions we take that are of no consequence. But there is another type of scenario where neural variability can be drawn on that is arguably more relevant for the question of free will and agent causality—it occurs not at the stage of choosing between options, but when suggesting them.

## Two-Stage Models of Free Will

In chapter six, we talked about options of what to do "springing to mind." In any given scenario, an animal has a limited set of ideas of what to do, based on its action repertoire, its past experience, its current goals, and its evolutionary imperatives. This limiting of options is the crucial payoff of all the learning the animal has been doing and is essential for rapid, efficient, optimal decision making. It means that the organism doesn't have to start from first principles, with an enormous search space of options, in every situation it finds itself in. But what determines *which options* spring to mind?

If the scenario is very familiar—like our morning routines—then one particular behavioral option may be so well worn as to have become habitual. In other cases, there may be a small set of previously used options that arise that are then adjudicated. (Think of your possible responses when someone says "good morning" to you.) So, the options we consider—the ones that even occur to us—are highly informed by our prior experience. But even in familiar scenarios, the particular ideas that bubble up to the surface may also be influenced by the inherent noisiness of the relevant neural circuits: they are unlikely to be *exactly the same* every time you are in any given situation. Indeed, given the variability we have been discussing, it's hard to see how they ever could be exactly the same in every detail. This notion is central to what is known as the two-stage model of free will, proposed by the American psychologist and philosopher William James in 1884 (see Figure 8.5). This model incorporates a degree of indeterminism in our cognition while protecting the causal role of the agent in actually deciding what to do. In the first stage, in any given situation, some set of possible actions occur to the organism. In this process, James proposed that some degree of randomness is at play, but the randomness does not decide the outcome—the organism does. The options are presented for consideration, and the organism selects the one that is most congruent with its current goals and beliefs and that has the highest predicted utility. That is, the organism selects from the range of presented options *based on its current reasons.*

In terms of our current understanding of the neuroscience of action selection as discussed in chapter six, the possible actions would be represented by patterns of activity arising in cortical areas. These patterns would then be evaluated through extended interlocking circuit loops among the cortex, basal ganglia, thalamus, and midbrain. This evaluation

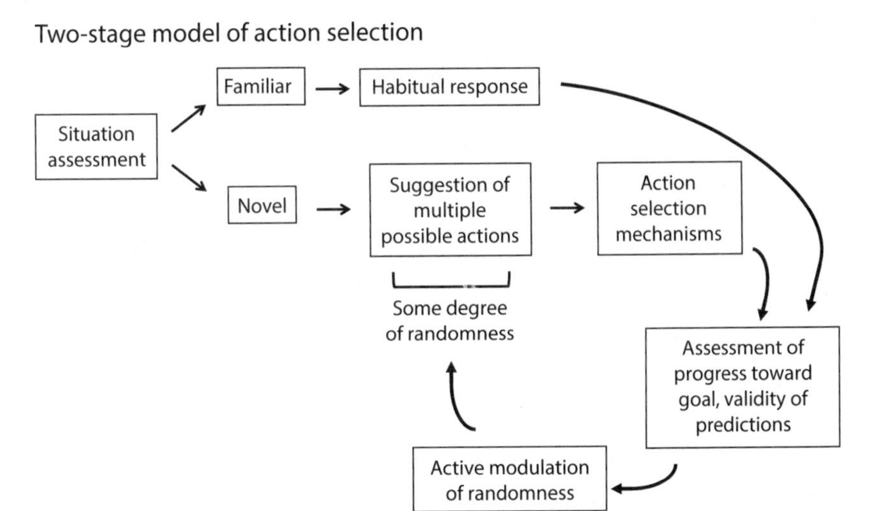


FIGURE 8.5. Two-stage model of action selection. If a situation is familiar, a habitual response may be "suggested" by the cortex, taken as optimal, and executed. In more novel situations, a range of options may become primed, with some degree of randomness in which ones happen to be activated (reflecting some ideas just "occurring to you"). They are then subjected to evaluation of the expected outcomes through the action selection circuitry, with one eventually being selected and released. In both familiar and novel situations, the outcome is monitored and evaluated with respect to the goals to see whether the behavior has been successful or remains optimal. When selected actions fail to achieve goals, signals may be sent back to the cortex to request additional options. The degree of randomness in that process may itself be modulated to increase the "search space" and find more creative solutions.

process biases the ongoing competition among the patterns of activity in the cortex, "up-voting" some and "down-voting" others, which ultimately results in one possible action winning the competition and being released while all the others remain inhibited.

This model thus powerfully breaks the bonds of determinism, incorporating true randomness into our cognitive processes while protecting the causal role of the agent itself in deciding what to do. As James put it, "Our thoughts come to us freely. Our actions go from us willfully."[4]

4. William James, The dilemma of determinism. *Unitarian Review* 32 (1884): 193. Reprinted in *The will to believe* (New York: Dover, 1956), 145.

James noted and was likely inspired by the parallel with evolution by natural selection in Darwin's *The Origins of Species*, published only a few decades earlier. In this process, new variations are generated by random mutation or sexual recombination but are then selected by natural selection. More recently, Raymond Noble and Denis Noble highlighted a similar dynamic at play in the immune system. When an organism is challenged by a new pathogen, a process of hypermutation is triggered in immune cells, rapidly changing the DNA sequence of genes encoding antibody proteins to generate novel antibodies. Immune cells expressing these antibodies are then exposed to selection by the immune system in the search for ones that match and can bind to antigens presented by the pathogen. Again, the process of generating novelty in this system is inherently random: the power comes from combining it with the subsequent selection process.

Importantly, in this model, it's not that *individual* random events at the quantum level decide what the organism does or generate new ideas. It's that the general randomness and thermal fluctuations cause a kind of variability in neural networks that can jostle them out of the ruts of habit and into potentially novel states. The influence of this variability will vary from situation to situation. In highly habitual behavior, there is no need to suggest lots of new options; indeed, it would be wasteful to do so. But there are many circumstances where the organism may need to actively draw on that randomness to think outside the box. This is especially true when an organism is in the process of learning and adapting to its environment, when it is in a scenario it hasn't encountered before, or when something changes and its prior strategies are no longer working. When there is no obviously better option or when none of the options is good, an organism may need to go back to the drawing board and come up with some genuinely new ideas. In many organisms there appear to be dedicated neural mechanisms that can control the degree of neural and behavioral variability in these various kinds of scenarios.

For example, in songbirds like the zebrafinch, juveniles learn their song from an adult tutor and practice reproducing it. To do so, they have to match their own motor output—the sounds they produce—to the learned template of the tutor's song. This takes some exploration of motor patterns, similar to babies babbling as they learn to talk: they figure out how to make different sounds through trial and error. Some randomness is valuable in this process as a means of tweaking the output and thus exploring the space of possible motor patterns. In this case, the degree of randomness is under the active control of a particular brain area, which is independent of the ones that actually do the learning. Neurons in this area fire in a highly variable way and seem to take a "copy" of the current motor pattern, add noise to it, and transmit it back to the circuits that evaluate how good a match it now produces to the template. When this area is inactivated, juvenile birds do not explore the space of motor patterns in the normal way, and their song crystallizes before it is a good match to the template.

Although this example is specific to songbirds, there are clear parallels to the tendency for juvenile animals of all kinds to explore through play, adapting through trial and error to the regularities of their particular environment. Indeed, some amount of noise is beneficial when neuronal networks are learning because it makes them more robust to variability in the environment and better able to generalize to new situations. In computing terms, it prevents the network from *overfitting* the possibly arbitrary features that it has experienced over any learning period. Again, there may be active systems that ensure this does not happen. Neuroscientist Erik Hoel has proposed that one function of dreaming may be to act as a kind of neural replay of recent experiences but with substantial randomness added, precisely to loosen the connections and associations formed by learning, some of which may not be genuinely predictive of future contingencies.

A similar system for regulating neural variability to tune up the exploratory nature of decision making seems to operate in mammals. When rodents or other mammals, including humans, are in a situation where their prior learned strategies are no longer paying off, neurons in a brainstem region called the locus coeruleus (or *blue spot*) become activated. These neurons project to many brain regions, including much of the forebrain, where they release the neuromodulator noradrenaline. In general, this signal increases arousal, prompting greater vigilance and readiness to act. But in one particular target region called the *anterior cingulate cortex* (ACC), it has the effect of increasing neural variability. The ACC is involved in monitoring action outcomes relative to goals and in setting the behavioral strategy: either to continue with a currently successful strategy or switch to something new. The noradrenaline signal increases the tendency to switch to something new. This may involve a constrained exploration of alternative actions that are still informed by the current model of the environment. But it can also involve the organism rejecting the model altogether—effectively throwing its hands up in the air and saying, "This just isn't working. I don't know what's going on!"—and exploring random options instead. This capacity to generate and then select among truly novel actions is clearly highly adaptive in a world that refuses to remain 100 percent predictable.

In humans, we recognize this capacity as *creativity*—in this case, creative problem solving. When we are frustrated in achieving our current goals or when none of the conceived options presents an adequate solution to the current problem, we can broaden our search beyond the obvious to consider new ideas. These do not spring from nowhere but often arise as *cognitive permutations*: by combining knowledge in new ways, by drawing abstract analogies with previously encountered problems in different domains, or by recognizing and questioning current assumptions that may be limiting the options that occur to us. In this way, humans become truly creative agents, using the freedom conferred by the underlying neural indeterminacy to generate genuinely original thoughts and ideas, which we then scrutinize to find the ones that actually solve the problem. Creative thoughts can thus be seen as acts of free will, facilitated by chance but filtered by choice. As dual Nobel Prize–winning chemist Linus Pauling said, "If you want to have good ideas you must have many ideas. Most of them will be wrong, and what you have to learn is which ones to throw away."[5]

5. As quoted by Francis Crick in his 1995 presentation, "The Impact of Linus Pauling on Molecular Biology," http://oregonstate.edu/dept/Special_Collections/subpages/ahp/1995symposium/crick.html. See also https://quotepark.com/quotes/1508216-linus-pauling-the-way-to-get-good-ideas-is-to-get-lots-of-ideas/.

## Rerunning Reality

Physicist Robert Doyle has highlighted how William James put chance in its proper place in the two-stage model of free will: "James was the first thinker to enunciate clearly a two-stage decision process, with chance in a present time of random alternatives, leading to a choice which selects one alternative and transforms an equivocal ambiguous future into an unalterable determined past. There are undetermined alternatives followed by adequately determined choices."[6]

There is a clear parallel between this framing and the models discussed in chapter seven, which conceive of the future as characterized by indefiniteness at both quantum and classical levels, and the present as the time *during which* this indefiniteness becomes definite. Note the crucial recognition in this framing that the present is not instantaneous: it has duration. The processes and events that resolve the indefiniteness take some time to play out. This is the period that we experience as "the thick present"—the period in which we are thinking of what to do and selecting among those options.

This brings us back to a thought experiment commonly discussed in debates about free will: whether, if you "rewound the tape" to put a person back in exactly the same circumstances they were in when they made some decision, they could, in this rerun of reality, do otherwise than they did the first time. Those who think the universe is entirely deterministic—that the physical state of the universe at that precise moment necessarily determines the state at the next moment, and so on—would, of course, answer, "No, they couldn't have done otherwise." As we discussed in the last chapter, even philosophers like Daniel Dennett, who think such determinism is compatible with the existence of a kind of free will, still accept that basic premise—in fact, the agent did not *really* have a choice at that imagined moment.

6. Robert O. Doyle, The two-stage model to the problem of free will: How behavioral freedom in lower animals has evolved to become free will in humans and higher animals, in *Is science compatible with free will? Exploring free will and consciousness in the light of quantum physics and neuroscience*, ed. Antoine Suarez and Peter Adams (New York: Springer Science+Business Media, 2016), 7.

The problem with this argument is that no such imagined "moment" exists. There is no such thing as a point in time of zero duration, when a system (or the universe as a whole) has a completely defined state: the Heisenberg uncertainty principle rules that out. In the present, the universe is still in a state of becoming, *in the process of* resolving all that indeterminacy, of the indefinite becoming definite. There's no point at which you can catch it with full precision until it becomes the past. By that point, of course, you could not have done otherwise because what's past is passed, but before that—*during the present*—things remain undetermined and real choice exists.

Martin Heisenberg—the son of Werner and an accomplished neuroscientist—noted in a 2009 essay that physical indeterminacy opens the door for true agency and free will. The constant jitteriness of neural activity means that the whole system is not predetermined to adopt any particular state: there are degrees of freedom in the system that the organism can exploit. The agent itself has both the power and the time to decide. Indeed, we have time to think, and choose, and change our minds, and think again if we need to. Even in the experiments of Libet and colleagues, where arbitrary noise is allowed to drive the decision of when to move, the subjects still retained the power to consciously veto that urge, up to just tens of milliseconds before execution of the movement.

As T.S. Eliot wrote in "The Love Song of J. Alfred Prufrock":

> *In a minute there is time / For decisions and revisions which a minute will reverse*

It is thus wrong to think of behavior as simply reflecting the inevitable transition of defined physical states of the brain from one instant of time to the next. No such states exist; no such instants exist. And there is nothing inevitable about the trajectory of neural states that the brain will follow through time. It is neither predetermined nor simply a passive response to stimuli from the environment. As Epicurus said, some things are caused by necessity, some are due to chance, and some are up to us. It's that *up-to-usness* that is the key element of agency. The elements of randomness at work in our brains give some leeway for us to have the final say in settling the matter.

# 9

# Meaning

In the last two chapters, we saw that there is fundamental indeterminacy in the physical world and that it manifests as a general level of molecular noise in neural circuits. The nervous system may directly draw on this underlying randomness to guide behavior in some scenarios, such as increasing the variability of exploratory actions during learning, expanding the range of action options to be considered when the current model of the world is no longer predictive, generating unpredictable actions to escape a predator, or even breaking a deadlock between arbitrary options.

But there is a far more fundamental and general implication of the inherent indeterminacy of the brain as a physical system: simply put, the momentary low-level details of atoms and molecules, or even the slightly higher-level details of firing of individual neurons, *do not determine the next state of the system.* This opens the door for higher-order features to exert causal power in "settling the outcome," as philosopher Helen Steward puts it; that is, deciding how the system evolves from moment to moment.[1] Indeed, it allows the whole system—the organism itself as a causal agent—to be in charge of its own behavior.

I argued in the preceding chapters that the higher-order features that guide behavior revolve around purpose, function, and meaning. The patterns of neural activity in the brain have meaning that derives from past experience, is grounded by the interactions of the organism with

1. Helen Steward, *A metaphysics for freedom* (Oxford: Oxford University Press, 2012).

its environment, and reflects the past causal influences of learning and natural selection. The physical structure of the nervous system captures those causal influences and embodies them as criteria to inform future action. What emerges is a structure that actively filters and selects patterns of neural activity based on higher-order functionalities and constraints. The conclusion—the correct way to think of the brain (or, perhaps better, the whole organism) is as a cognitive system, with an architecture that functionally operates on representations of things like beliefs, desires, goals, and intentions.

But is that view really justified? Are beliefs and goals and desires and intentions really doing the causal work in the brain? After all, I described how these cognitive elements are encoded—literally instantiated—in the physical structures of the brain and the resulting patterns of neural activity. Indeed, modern neuroscience is doing an awe-inspiring job of revealing the underlying neural mechanisms of perception, decision making, and action selection. We can see from incredibly powerful experiments in animals how driving particular patterns of activity in certain neural circuits can cause the animals to perform certain actions: move, sleep, eat, fight, freeze, mate, hunt, turn right, turn left, stand up, roll over. We can even selectively and subtly manipulate the animal's cognitive processes: changing its reward sensitivity, its confidence threshold, or the weighting of different goals. We can even implant memories and instill false beliefs.

These experiments clearly show the causal power of neural activity in driving the cognition and behavior of the animal. So much so that you might ask whether there's actually anything for the supposed *semantic content* of these patterns of activity to do. Does it even matter what these patterns *mean*? It seems to be the case that if you just drive these circuits or those ones, then the animal will behave like so or think like so. This encourages a view of the nervous system as a complex machine, with the causal power vested in its separable internal mechanisms. It is summed up in a quote from Francis Crick, who, after co-discovering the structure of DNA, went on to pioneer the study of consciousness as a problem for experimental neuroscience: "'You,' your joys and your sorrows, your memories and your ambitions, your sense of personal identity and free will, are in fact no more than the behavior of a vast assembly of nerve cells and their associated molecules."[2]

It's clear that all those psychological attributes and functions *depend on* the behavior of a vast assembly of nerve cells and their associated molecules, but are they really "no more than" that? We may have escaped the reduction of agency to the elementary particles and forces of physics, but can we escape a reduction to neural mechanisms? In our attempts to explain the neural basis of cognition, will we in fact explain it away? There is certainly a strong implication that beliefs and desires and intentions are mere epiphenomena that come along for the ride with certain patterns of neural activity, without themselves doing anything. In this chapter, I argue for precisely the opposite view—that the neural patterns *only have causal power in the system by virtue of what they mean*. The underlying neural mechanisms are simply how those meanings are instantiated. By looking at the details of communication among neurons, we can see how meaning is extracted or even created in the nervous system.

## Black Boxes

Recall that one of the key properties of living organisms is that they are causally insulated from their surroundings. They are not in thermodynamic equilibrium with their environment; they hold themselves apart and work hard to maintain that separateness. Things happen chemically and physically in their surroundings that do not have any effect on the inner goings-on: organisms are sheltered from the storm outside by their boundaries with the environment. But they are not indifferent to what's happening out there. They actively sample information about relevant factors out in the world. For single-celled organisms, this mainly means various chemicals that mark out food or fellow critters or threats. Each type of single-celled organism can only sniff out a tiny fraction of the chemicals around it while remaining oblivious to everything else.

2. Francis Crick, *The astonishing hypothesis: The scientific search for the soul* (New York: Scribner, 1994), 3.

Cells in our body are like that too. They are only aware of the highly selective chemical information to which they are each attuned. For neurons, this means the neurotransmitter molecules and other signaling factors (neuromodulators and hormones) secreted by other cells. For each neuron, the other neurons it is connected to remain black boxes. It has no access to what is going on inside them; the only information it receives is the occasional presence of neurotransmitter molecules that get bound by the receptors at the receiving side of all the synapses made with it by those other neurons.

When neurons communicate with each other, the point is not merely to transmit faithfully a signal from one neuron to the next; they are not like phones that merely encode, transmit, and then decode or reproduce some signal. In fact, it's not always obvious what "the signal" is or how it is encoded. When neurons receive inputs, their neurotransmitter receptors get activated, ion channels open, and sodium ions rush into the dendrite, the branch of the neuron where a given synapse is located. This creates a graded change in the electrical potential across the membrane, but initially the change is very local, near the synapse. Similar changes happening across the dendritic tree may peter out, or they can feed into the cell body of the neuron. If they collectively push the potential above a threshold, then an amplification process occurs that triggers a spike: an electrical impulse that travels down the axon to the synapses onto other neurons. The signal is thus converted from an analog code (graded potentials in the dendrites) to a digital code (a discrete spike or series of spikes) and then back again at the next synapse.

We can see multiple important principles at play here. First, detailed information on spike timing is usually lost in synaptic transmission. There are some exceptions, for example in the auditory system, where high-frequency signals need to be transmitted faithfully because the signals are about high-frequency stimuli. But most neurons are not set up to respond completely faithfully, spike for spike, to individual inputs. Instead, their job is to integrate and transform information by summing inputs, often over multiple synapses and over some time window. The threshold nature of spike initiation means that it is a highly nonlinear process; that is, the output does not depend linearly on the inputs. If the collective inputs are below threshold, there is no change in the output. And if they are above threshold, there is a unitary change—a spike occurs. There is thus an all-or-nothing output at a given moment. However, over some period of time, a neuron that is driven more by its inputs will fire *at a greater rate*. That firing rate, rather than a precise pattern of spikes, is typically what other neurons are listening to.

This means that most of the low-level details of what is going on inside each neuron are lost in transmission. The other neurons don't have access to those details, and they're set up not to care about them. Most of the time, they are only sensitive to the firing rate, which is where the information lies. This is an example of what is referred to as *coarse graining* of information, the idea being that the very fine details at the low levels are averaged over to give a coarser description at a higher level. However, this term is a little unfortunate, in that it seems to imply only a loss of information in the transformation from fine grained to coarse grained. This is not the case. The details may be lost, but *a new type of information* is gained in that process, even in transmission from one single neuron to another. The first neuron doesn't "know" what its firing rate is; it is either currently firing a spike or not. It takes another neuron to monitor the firing rate by integrating spikes over some period of time, creating a new kind of information in the process.

## Beyond Coarse Graining

When we put lots of neurons together in circuits, they can generate all kinds of new information. Because each neuron typically integrates inputs from many other neurons, they are set up to perform diverse logical operations on the various types of information carried by these inputs. Rather than thinking of the receiving neuron as a passive element in this process, simply being driven by its inputs, we can think of it as actively monitoring its inputs and integrating that information to settle on its new state of activity.

The fact that simple neural circuit motifs can perform logical operations was noted in a famous paper in 1943 by Warren McCulloch and Walter Pitts, which laid the foundation for the field of artificial intelligence. The key point in this idea is that *the wiring sets the criteria for firing.* For example, imagine two neurons, A and B, that both input to a third neuron, C. There are many ways to set the wiring so that the firing of C will mean different things. If, for example, both A and B make very strong synapses onto C, so that activity of either of those inputs is sufficient to drive C to fire, then the firing of C means "A OR B." (That's shorthand for saying that either A is firing or B is firing, which in turn means that whatever neuron A's firing means is the case or whatever neuron B's firing means is the case.) If A and B both make weak synapses onto C, so that it takes both of them to activate C, then C's firing means that "A AND B" are the case. And if, say, the synapse from B is inhibitory instead of excitatory, so that it can cancel out activation from A, then C's firing would mean "A NOT B" (that is, A is active and B is not).

By surveying its inputs in these various ways, the firing of neuron C creates new higher-order information. You might argue that the information about the conjoint activation of A and B is in principle available in the firing rates of A and B, but in practice the information pattern does not exist anywhere until C creates it by actively making the comparison between A and B *over a given period of time.* If we just have the firing patterns of A and B independently and we do not know how they align in time, then we cannot extract any higher-order relationships. It thus requires a higher-order structure to be in place, as well as some amount of work to be done, to create this higher-order information and make it available to the rest of the nervous system.

We already saw these kinds of operations at work in the circuitry of the retina and the extended visual areas in the cortex. Retinal ganglion cells (RGCs) integrate information from many neighboring photoreceptors and make comparisons between them to extract higher-order features such as contrast between neighboring spots of the visual field, for example. That information is not available in the firing of the photoreceptors alone—it requires knowing *where they are relative to each other.* If RGCs only integrated inputs from photoreceptors randomly placed around the retina, they couldn't make these kinds of inferences. It is the physical configuration of the circuitry that allows their activity to encode some new information that is useful to the organism. The same is even more true for color perception, which depends entirely on a comparison across photoreceptors responsive to different frequencies. This is a clear case of the *creation* of information—of new meaning—that does not exist at all in the individual firing patterns of the photoreceptors. Indeed, color doesn't even exist in the world: it is the result of the organism creating what are essentially arbitrary categories to help distinguish objects from each other.

The logic for populations of neurons is very much the same. We saw in the last chapter how populations can exhibit quite restricted trajectories through *state space*. There are often a small number of states that the whole population can stably occupy because of the complex network of feedforward and feedback connections, both excitatory and inhibitory. These states will come to encode or represent something, depending on these factors—the pattern of inputs that population receives, the information they carry, the operations performed on those inputs by the receiving neurons, and the criteria embodied in those connections—that determine the response of the receiving population to the different possible patterns of inputs.

In the visual system, for example, higher-order features like faces are encoded at the population level in high-level areas of the hierarchy of visual cortical regions. A certain type of activation in that region *means that* a face has been detected in the visual scene. The activation in the face-processing region can then be made available to many other parts of the brain—effectively acting as a report that a face is out there in the world. This information is formally present in the patterns at lower levels of the hierarchy, but it is so entangled with all the other information that it is not accessible without this extra processing. Links to memory systems in other parts of the cortex allow recognition of finer patterns in the "face area" as corresponding to the face of a parent, or a friend, or a stranger, possibly triggering all kinds of thoughts and feelings and behaviors.

This kind of neural coding is essential for the system to do its job efficiently, both energetically and computationally. Because sending spikes is really energetically expensive, there are strong selective pressures to compute as much as possible locally and to send as little information as possible over longer distances. This means that extracting the most meaningful information and leaving the irrelevant details behind are highly favored strategies. This can be done by abstracting or disentangling higher-order features and making them the objects that the rest of the system operates on.

This is an essential principle in computer science and linguistics. If you want a program to be able to operate on something—to take its value as an input and perform some kind of logical operation on it to derive an output—then you have to *name it*. For example, in data that have some hierarchical structure, identifying and naming higher-order categories enables the program to refer to these categories as variables, to know where to find the information about them, and to call that information up when needed. The program can now keep track of and perform operations on the higher-order categorical objects that it could not perform directly on the low-level data.

In our own languages, naming things in this way gives access to that term as an element in recursive thought. Imagine how cumbersome it would be if you had to use long descriptive terms for groups of things—like, for example, those furry things with four legs, and a tail, sharp teeth, about this big, that like to hang around humans. That would make communication exhaustingly inefficient, but it would also make internal thought so time consuming and imprecise as to hardly be worthwhile. You couldn't really perform operations on those nebulous descriptions: they are not concretized into a thing. They are just a bunch of properties that individual members might embody better than others. You cannot think about dogs, in other words, without having a concept labeled "dog." When you name a category like that, it becomes *a thing*. You are recognizing its thingness and treating it as such. The name is not just a handy label—it reifies a concept and makes it a cognitive object.

Shaping cognition in this way, framing it around categories with defining properties, allows organisms to make predictions about these categories, identify causal relations between specific things and between types of things, and even recognize types of relations. These abilities lead to what we might take to be the hallmark of *understanding*: the ability to generalize to new situations by reference to an organized hierarchical, categorical framework of regular causal relations in the world. We will see in the rest of the book how naming cognitive objects enabled the system to develop a capacity for self-reflection and metacognition: to think about its own thinking.

Encoding meaning in patterns of neural activity is thus efficient and cognitively powerful. In addition, it is essential if the nervous system is to do its job robustly. If neural circuits and networks were sensitive to the minutiae of every change in electrical potential, every nerve impulse, or synaptic event, they would be extraordinarily vulnerable to even small environmental fluctuations, intrinsic noise, and the momentary failure or loss of individual components. Encoding information over brief windows of time across ensembles of neurons buffers these sources of variability and ensures that only meaningful changes affect the evolution of the system.

This is the norm at least, although there are also those cases we considered in the last chapter where the nonlinear dynamics of some circuits allow intrinsic variability to be actively put to use. In addition, by sampling over time, any variability in neural activity—for example, a population oscillating between two possible states—itself becomes a source of valuable information. When assessed by downstream neurons, such variability can be used as a meta-signal of uncertainty, allowing a population to convey both its best guess about what its current inputs mean and indicate how much weight should be put on that assessment.

## Grounding Meaning

The meaning of internal signals has to be grounded in some way. For a pattern of neural activity to be meaningful to an organism, it should refer to something—that is, be physically correlated with and thus carry information *about that something*—and it should be consequential, at least potentially. If nothing would ever change as a result of that information—the fact that the pattern is one way, when it could have been some other way—then it's not really meaningful: instead, it's functionally irrelevant.

We saw in earlier chapters how the first glimmers of meaning arise in relation to the primary goal of the simplest organisms—to persist. In unicellular organisms, sensory signals are often coupled closely to motor responses (with some integration of other internal and external signals). And natural selection is the critic that judges whether a certain response is good or bad for the persistence of the organism. The meaning in this system is thus not in the sensory signal itself: it resides in the control policy governing the pragmatic response to that signal and the effect that has on the organism's survival.

In multicellular organisms with nervous systems, sensory neurons respond selectively to particular types of stimuli: the binding of an odorant molecule, an air vibration of a certain frequency, the detection of a photon refracted through the lens from a particular point in the visual world. The activity of those primary sensory neurons thus *correlates with* and carries meaningful information about those physical things in the world. Importantly, these kinds of primary signals do not have to be external to the organism—they can be reporting the state of the body itself as well. Similarly, in the outward direction, the activity of neurons controlling motor action plans means "execute this series of muscle movements in order to act on the world." Again, there are direct physical correlates of the patterns of activity.

But direct pragmatic couplings only get you so far. It's difficult to build a system capable of integrating multiple signals and coordinating multiple subsystems in that purely mechanical way. There are only so many cogs you can connect to each other without jamming the whole mechanism. Increasingly sophisticated behavior required decoupling sensory signals from motor output. In multicellular organisms, this led to the evolution of intervening layers of neurons and the emergence of *internal representations*. The meaning of these representations has become semantic, decoupled from an obligatory physical response and the value of its outcomes.

You might naturally ask, How does the firing of some internal neuron A or the pattern of population B *come to mean some particular thing*? We don't have to fear some kind of infinite regress here, because we still have the boundaries of the organism and its interactions with its environment to ground the meaning. The meaning of the patterns of each new intervening layer of interneurons is grounded by the inputs on the sensory side and by the eventual outputs on the motor side. The hierarchical circuits that interpret sensory data generate inferences of higher-order features out in the world, which is what the neural patterns correlate with. Their activity thus represents *the belief that* there is some higher-order feature out in the world (as we saw, such beliefs can be mistaken). Similarly, areas representing internal states can report on hunger or tiredness or thirst and thus provide motivation for the adoption of different *goals*, which are also internally represented. Areas involved in motor planning (one level removed from execution) can represent *the intention* to execute some motor action sequence, without the organism having to actually pull the trigger. Ultimately, these representations are all incorporated into models of the world, models of the self in the world, models of potential actions, and models of their potential value. All these things *collectively ground the meaning* of all the patterns (see Figure 9.1).

## Mapping Meaning

This collective grounding of meaning has a concrete substrate in the structural connectivity of the brain. A major organizing principle in brain function is that different kinds of parameters are mapped in a smooth and organized way across the extent of different brain regions. We already saw some examples, such as the continuous maps of the visual world created in the tectum or in early levels of the visual cortex. Basic neurodevelopmental processes ensure that axons from the retina maintain their nearest-neighbor relationships when they connect with their targets in the brain. This creates a visual *topographic map*, with nearby neurons in those brain areas responding to stimuli from neighboring points in visual space. The same mechanisms ensure these maps are projected to subsequent recipient areas.

The visual example is the most concrete and easy to appreciate, but actually the principle of maintaining topographic relationships in the projections from one brain area to another is a general anatomical rule

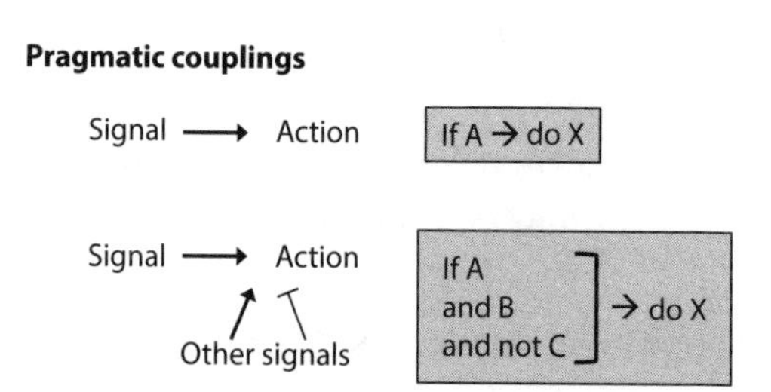


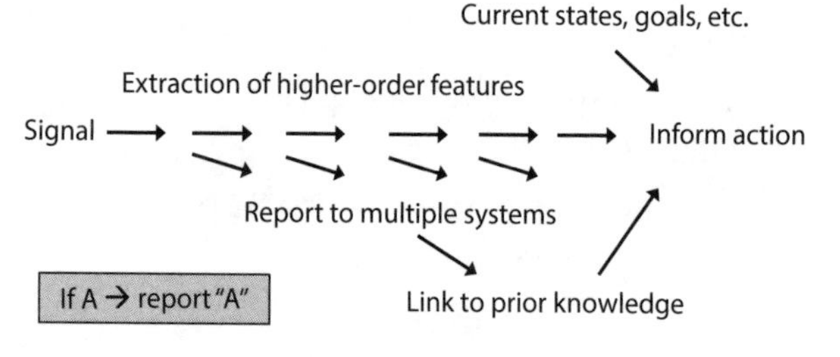


FIGURE 9.1. Pragmatic and semantic meaning. **(Top)** An internal signal (which could reflect some external parameter) may be linked tightly through a configured control policy to an action. The meaning of the signal A is thus "Do X." Additional signals (B, C) may be integrated at the same level to perform more complex, context-sensitive operations. **(Bottom)** With additional layers of internal processing, signals may be processed and reported to the rest of the brain without necessarily driving action. The meaning of these internalized representations is grounded through the web of associations accumulated by the organism and may still be eventually used to inform action.

in the brain. This principle is both developmentally and energetically efficient in that it minimizes the number of genes required to specify the connectivity and the length of wires required to connect neurons that are processing similar parameters or types of information. This not only saves on space and expensive building materials but also ensures the shortest distances and thus the fastest and cheapest communication for local computations.

This principle extends to the alignment of different kinds of maps to each other. For example, in the auditory system, topographic projections from the cochlea generate a smooth map of pitch across the auditory cortex. Comparison of the intensity of the signal of different pitches coming from the two ears then allows inference of the locations from which different sounds are emanating. The auditory system can thus also generate a map of objects in the world, which is directly aligned with that from the visual system in different layers of the tectum. In turn, these maps of objects in the world are aligned with maps of action that direct visual gaze or motion toward various parts of space.

All kinds of other things can end up being mapped in this way across brain regions. There are maps of broad families of chemicals in the olfactory system, maps of kinds of actions in the motor cortex, maps of short- and longer-term goals in the premotor and prefrontal cortex, and maps of navigational space and heading (which way an animal is facing or traveling) in the hippocampal system. And there are maps of concepts too—semantic categories that are represented in stereotyped positions and arrangements across individuals.

For example, in higher areas of the visual system where object identity is extracted and represented, there is a systematic map of different *kinds of objects* that is remarkably consistent across individuals and even between humans and monkeys (see Figure 9.2). This mapping is categorical and hierarchical, with similar types of objects represented by more similar patterns of activity in nearby clusters. There is, for example, a broad division of clusters representing animate versus inanimate objects. Within those clusters, there are subclusters distinguishing humans versus other animals and, at a finer level, human faces versus other body parts.

Remarkably, this mapping seems to extend across the border into nearby regions that mediate language, with an alignment between visual and linguistic representations of different types of objects. There thus appears to be an anatomical progression from areas representing percepts to ones representing concepts. The map of concepts also receives inputs from memory pathways encoding the schemas or webs of associations that define our conceptions of each type of object. Studies of bilingual individuals found very similar brain responses in these regions to hearing words in their two languages that mean the same thing but have completely different phonetic properties. By acting as convergence zones for these different types of information, as well as for signals

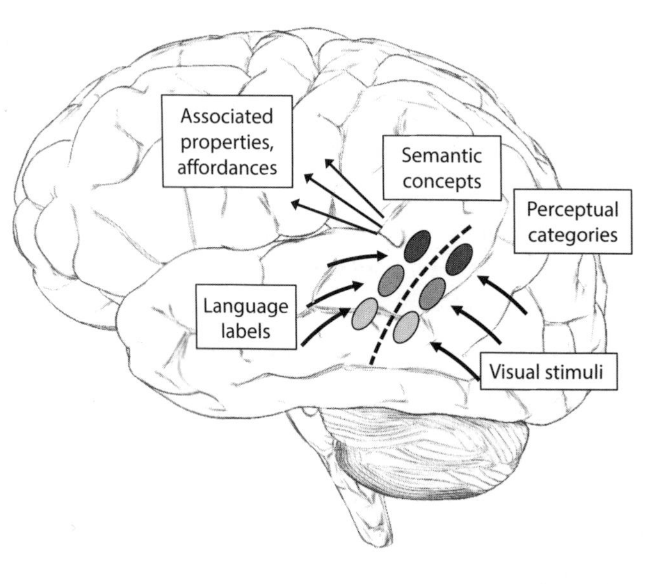


IGURE 9.2. Maps of semantic categories. Sensory information (shown here for the visual system) is parsed and processed to extract features and objects of relevance to the organism. Higher levels of processing can come to represent objects in a disentangled and invariant way, allowing them to be recognized as object A even from different views or in different contexts, for example. Yet higher levels can represent categories and types. Representations of such concepts are mapped across these higher-level visual regions in a stereotyped way across individuals—with animate and inanimate objects represented in different parts of the map, for example, and additional distinctions between humans and animals, faces and body parts, and so on. Remarkably, the same organization is observed in neighboring areas representing linguistic categories. This suggests an overarching mapping of high-level semantic concepts, independent of specific sensory information.

representing emotion and valence, regions of the temporal and parietal lobes thus come to encode a structured map of abstract semantic space. Notably, the number and territory devoted to these so-called association areas of the cortex are particularly increased in human evolution.

Meaning in these systems is inherently *relational*: the meaning for the organism of any given pattern of neural activity representing X inheres in the relationships that X has to Y, and Z, and so on. These relationships are embodied in a highly structured web of connections capturing

categorical and hierarchical information. This struct
of cognitive navigation, of moving mentally from ide
ing connections and tracing relations. Indeed, it seem
tive processes are supported by the same mechanism
navigation and memory in the hippocampus. As a g
then, the topology of neural connections is highly stru
abstract meaning to be grounded in a convergence of per
ties and stored knowledge of relations, and providing a na
for a higher-order cognitive map of concepts.

## A Note on Neural versus Mental Representa

I should explain that my use of the term "representation" he
necessarily imply that the representations are *mental*. They
but I am referring to *neural* representations: patterns of neura
that correlate with things in the world or that correspond to t
of cognitive elements discussed earlier, such as beliefs, goals, int
and so on. Admittedly, those sound like inherently mental phen
that should necessarily be experienced in the mind, but we can r
agnostic about that for now. In fact, these kinds of functional
could be instantiated in the parameters of the control system of a r
that, for all we know, might not be experiencing anything at all. Thu
theory at least, these cognitive elements can similarly be instantia
in the nervous system without necessarily manifesting as an expe
enced thought or an idea or a mental image.

We have not spoken much about "the mind" so far, partly because th
concept is so poorly defined and encumbered by a historical load o
dualist baggage. It is common to think of the mind as an object somehow separate from the brain; a private arena populated by immaterial thoughts and ideas. However, this raises the classical problem of how the immaterial stuff or processes of the mind could intervene on the physical stuff and processes of the brain. In reality, mind and brain cannot be separated like this. A more accurate conception of the mind is as an interlocking system of cognitive activities that are necessarily mediated by the functions of the brain. In humans, some of those cognitive

activities are associated with conscious mental experience, but they don't all have to be to be effective.

Thinking too much from a human-centric point of view can bias our intuitions here. The problem, of course, is that human minds are the only ones to which we have firsthand access. We don't have a good idea what the minds of other animals could be like, if they can be said to have them at all. I personally do not see any good reason to deny that animals have their own kinds of minds; when we regard different kinds of creatures, there certainly seems to be "something going on in there" for many of them. But it's impossible to know from our secondhand perspective what their subjective experience is like. However, we can get some idea of the limits of what subjective experience *might be about* in various animals. An animal cannot be "thinking" about things like distant objects or maps of things in space or narrative sequences of events or imagined futures if they lack the neural architecture to create that type of information and represent those kinds of meanings.

What is interesting to note, in our own case at least, is what is excluded from subjective experience. We don't experience the firing of our neurons or the flux of ions or the release and detection of neurotransmitters. *What we do experience is what patterns of neural activity mean,* at the level that is most relevant and useful and actionable for the organism as a whole. In vision, for example, we don't experience individual photoreceptors absorbing photons of light or individual RGCs comparing inputs to create information about contrast. What we experience is a map of objects out in the world, relative to each other and to ourselves, derived not just from vision but also from our other senses. And we ourselves are not absent from that experience: our perception of our own selves, our own bodies and minds, is the ever-present background context for all these perceptions. We experience *being ourselves experiencing things.*

We will return in later chapters to the distinction between conscious and subconscious processes and the question of what are the benefits that consciousness and reflective experience can bring to agents making their way in the world. For now, the important point to remember is that cognitive elements like beliefs, desires, and intentions can be encoded in patterns of neural activity that mean something for the organism and that it is those meanings that inform and guide our choices.

## Meaning Drives the Mechanism

This brings us to an absolutely crucial property of the coding of information in neural circuits, known as *multiple realizability*. That awkward-sounding phrase simply means that a given meaning can be realized physically in multiple ways. Or, conversely, many different low-level arrangements (microstates) can correspond to the same higher-level pattern (macrostate). We're familiar with this idea from language: the words on this page could be written in a different font or in braille, or could be read in an audiobook in various accents, and they would always mean the same thing. In other words, the low-level details of how the meaning is instantiated don't matter. Similarly, what matters in the nervous system is the larger-scale pattern: the firing rate of individual neurons or the low-dimensional collective state of neural populations.

This fact is key to countering an otherwise stubbornly reductive and overly mechanistic view of neural activity. We saw that physical indeterminacy means that the current physical state of some system does not fully predict the next state. In the brain, the resultant noisiness at the molecular level can have an impact on neural firing and affect the evolution of patterns of neural activity. That seems to free us from straightforward physical predeterminism. But you might argue that in such a scenario the way that neural activity evolves through time is still completely driven by the low-level physical parameters of the system, even if some of them are noisy. This argument would maintain a reductionist view in which all causation comes from the bottom up, even if strict determinism does not hold. If that argument were true, then changing the low-level details in one neuron or one population at a given time should necessarily change the firing parameters of the neurons or populations it connects to.

Multiple realizability shows this is not always the case. There are many possible changes to the detailed parameters of firing in a neuron or a population that do not have any impact on downstream targets (see

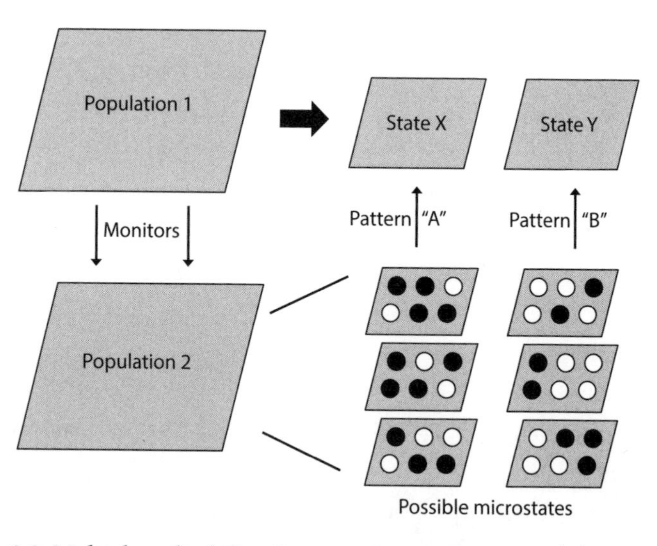


FIGURE 9.3. Multiple realizability. One population of neurons (1) is shown monitoring the activity of another population (2). Population 2 may exhibit a diverse set of possible microstates, each defined by the activity of all the neurons in the population. However, these microstates may correspond to either macrostate A or B from the perspective of Population 1. The response of Population 1 (whether it adopts state X or state Y) may thus depend on the overall macrostate (and its meaning) and not on the low-level details of the microstates in Population 2. In this way, the meaning of the neural patterns is what drives the mechanism.

Figure 9.3). For example, if neuron 1 is monitoring inputs from neuron 2, and it cares about the rate but doesn't track the precise timing of every spike, just summing excitation over some duration, then many different timings of spikes will be equivalent in terms of the information that the whole train of impulses carries (because neuron 1 can't tell the difference and doesn't care). Similarly, if population 1 is monitoring inputs from population 2, there may be many details of those inputs that are irrelevant because the dynamics of population 1 will tend to cause it to settle into one of its *stable attractor states,* and those initial details will have been lost in the process. Population 1 cares about the macrostate of its inputs, not the particular microstate.

If changes to the low-level details have an impact in some cases and not others, then simple reductionism does not hold. Something else determines what separates these cases, and that something else is the way the system is organized. The brain doesn't run on spikes but on patterns. The way in which system activity evolves through time is driven *by what those patterns mean*. That meaning is grounded in the history of interactions of the organism with its environment and is embodied in the patterns of synaptic connectivity between individual neurons or functionally distinct populations. Those connections set the criteria for how each receiver will interpret and respond to the incoming information.

Neuroscientist Peter Ulric Tse refers to this dynamic as "criterial causation." Although it is necessarily instantiated in a set of physical mechanisms, the causation in the system is informational. The criteria for how each neuron or each population responds to incoming information are set over multiple timeframes: over millennia by evolution; over a lifetime by individual experience; over years, months, days, or hours by the adoption of different goals; and over minutes, seconds, or even tens of milliseconds by processes of attention, arousal, and moment-to-moment decision making. The choices the organism makes based on parameters at a high level filter down—sometimes extremely rapidly—to change the criteria at lower levels, thereby allowing the organism to adapt to current circumstances, execute current plans, and achieve current goals.

In this way, abstract entities like thoughts and beliefs and desires can have causal influence in a physical system. The idea that they are mere epiphenomena that just come along for the ride, while neural mechanisms do all the real work, could not be further from the truth. Patterns of neural activity only have causal efficacy in the brain *by virtue of what they mean*. Of course these higher-order abstract entities must be instantiated in a physical medium, but they are not reducible to the underlying physical mechanisms. Nor do they magically just emerge from those mechanisms. The meaning of the various (often arbitrary) neural patterns arises through the grounded interaction of the organism with its environment over time (see Figure 9.4).

There is nothing mystical or vague or dualist about this idea of top-down or semantic causation, nor does it violate physical law. In fact, we have an utterly commonplace example of it in our everyday lives in

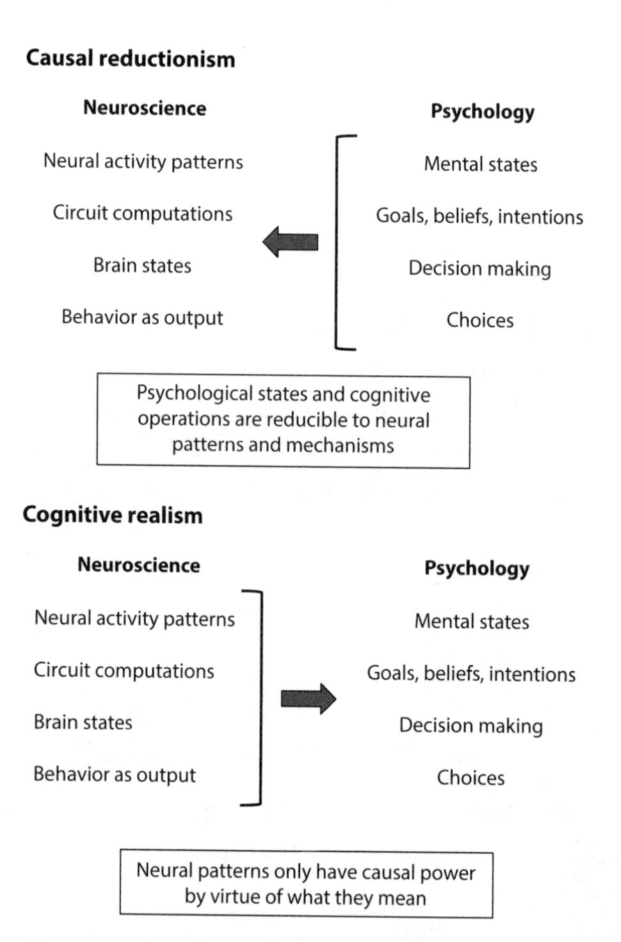


FIGURE 9.4. Meaning drives the mechanism. Under *causal reductionism*, psychological states and cognitive operations can be reduced to being "nothing more than" the activity of neural circuits, which may, in turn, be nothing more than the playing out of physical forces between molecules. The alternative view, which we could call *cognitive realism*, argues the converse—that neural patterns have causal power in the system solely by virtue of what they mean; that is, by virtue of their status as representing goals, beliefs, intentions, or other elements of cognition (whether conscious or not).

computer programs, as described by philosopher and mathematician George Ellis. Computer programs are abstract entities that can be instantiated in many different physical forms and even run on many different kinds of hardware, from a Mac to a PC to a smartphone, or even on much simpler computing machines using vacuum tubes or mechanical cogs. Once instantiated in some physical medium, a given program imposes a particular set of top-down constraints on the behavior of the physical components of the system.

The distribution or flow of electrons through the circuits of the computer are constrained so as to represent the values of variables that are operated on by the algorithms specified by the abstract program. The program dictates the flow of physical states by setting the configuration of the system, but it does not specify the precise low-level details required to execute how it runs. Any given element of the program or the data necessarily has some particular physical instantiation, but it is one arbitrary instantiation of many that are possible, which all correspond to the same value or operation. If any elements of this process violated physical law or causality, there would be no computers. The fact that they exist and that they demonstrably work in this way clearly shows that there is nothing outlandish or fantastical about top-down or semantic causation.

## Justifying Agent Causation

The picture we have built up so far frames the living organism as a causal agent capable of acting in the world, exercising choice, and doing things for its own reasons. It's worth a recap here to make explicit the properties that justify this position. To say that a living organism is a cause of things—to locate causation at the level of the agent itself—requires that two general criteria be met.

First, the organism should be causally insulated from its environment. If it were not, it would in fact not be an organism at all, never mind an agent. To even talk sensibly about it being a cause of anything, it first has to be an identifiable physical entity that has some persistence through time, with an inside and an outside and a barrier between them. We saw how organisms actively maintain themselves not only by replenishing the material they are made of but, more importantly, also by constraining the relations between their components to yield a specific, self-sustaining pattern. That interlocking set of constraints is causally self-contained. The system doesn't require any outside direction to keep it that way; indeed it actively resists external perturbations by doing work to maintain its internal organization. Organisms thus have very substantial causal integrity and autonomy. This autonomy is not absolute of course, but organisms are not simply open to the environment and not just pushed around by it: in fact, they're capable of pushing back.

The philosopher Immanuel Kant put it well: "An organized being is, therefore, not a mere machine. For a machine has solely motive power, whereas an organized being possesses inherent formative power, and such, moreover, as it can impart to material devoid of it—material which it organizes. . . . This, therefore, is a self-propagating formative power, which cannot be explained by the capacity of movement alone, that is to say, by mechanism."[3]

Second, to say that the organism itself is doing things, what happens should be driven by the organism as a whole—*as a self*—and not just by the physical workings of some internal mechanisms. The organism should be in charge of its parts, not the other way around. Otherwise, it's not really an integrated system acting as an agent but just a place where complicated things happen. Kant continued, "The definition of an organic body is that it is a body, every part of which is there for the sake of the other (reciprocally as end, at the same time, means)."[4]

Thus, it is not really correct to think that the various components of an organism are parts, in the way we think of the separable parts of a machine. Instead, *they play parts*. It is their functional relations to each other that comprise the pattern of organization. We often draw simplified diagrams of systems with various elements—A, B, C, and so on—in little boxes and arrows drawn between them. The key point to realize is that the organism is not made of the boxes: it's made of the arrows. It is not an organization of stuff; it's an organization *of processes* in mutual relation to each other. The activity of each process is constrained by the collective activities of all the other processes, and these collective

3. Immanuel Kant (1790), in *The Critique of Judgement*, trans. J. C. Meredith (Oxford: Clarendon Press, 1980), cited by A. Juarrero Roqué, Self-organization: Kant's concept of teleology and modern chemistry, *Review of Metaphysics* 39, no. 1 (1985): 107–35.

4. Immanuel Kant, *Opus postumum*, trans. E. Forster and M. Rosen (Cambridge: Cambridge University Press, 1993), 64.

activities are constrained to be goal directed, with functionalities that are fitted to the environment, designed to ensure the thriving of the organism and its lineage in its environment.

With those general criteria of autonomy and wholeness met, we can look at more detailed properties that together support claims of agent causation versus a reductive, mechanistic, and instantaneous view.

First, decision making requires that multiple distributed subsystems act in parallel, communicating with each other over some period of time, through a complex web of interlocking, recursive circuits. Each area informs and constrains the others, until the whole system resolves ambiguities and conflicts and comes to some kind of consensus on what is out in the world, what its present goals should be, and what the best actions are to pursue those goals. In this way, the entire system *collectively* settles into a new state.

Note how different this scheme is from the workings of a machine with separable components or the linear series of algorithmic steps carried out by a digital computer executing a program. A living system cannot be deconstructed into constituent mechanisms, because their functionalities are inherently conditional on each other and thus inextricably intertwined. Indeed, it is becoming more and more clear that information about many behavioral parameters is widely shared across brain areas previously thought to work in isolation. Even early sensory areas typically display—in addition to the specific sensory information they are ostensibly processing—patterns of neural activity that correlate with information from other senses, or current motion, or tasks, or goals. Information processing is not done through completely separate channels and fed upstream to a central decision maker. Functions are still specialized and localized, but there is substantial cross-talk at all levels, and that global contextual information modifies signaling in each area.

To be clear, the proper operation of the entire system relies on the integrity of all its components and their proper functions. When some components are damaged or altered or manipulated, the behavior of the whole system can be affected, often in quite specific ways. Understanding how this happens is obviously of huge importance clinically; for example, in elucidating the impact of mutations in specific genes or damage to specific brain areas. As scientists we take advantage of this reliance on individual components to try and work out the logic of how the whole system works. But the fact that our experimental approach is often reductive in nature—manipulating single components while trying to control for other variables—should not bring with it a commitment to the idea that the system itself is reducible to its components. Once you break apart the relations, there is no system left to understand.

Second, organisms are endogenously active. They do not passively wait for stimuli to respond to. Even when they are physically still, they are not internally static. When signals come in from the sensory periphery, they are assimilated into the ongoing flux of biochemical and neural activity. For example, when neuroscientists perform brain scans on people using functional magnetic resonance imaging, the signals that are associated with the person doing some particular task are tiny: only about 1 to 2 percent the magnitude of the background bustle of neural activity. The brain is not just sitting there, waiting for signals. And, of course, most animals spend much of their time busily interacting with their environment—exploring, sensing, choosing, and actively shaping it. Both internally and externally, organisms are proactive, not just reactive.

Third, as we saw in this chapter, the currency of the nervous system is meaning: that's what causally drives the mechanism. This is not just information in an abstract mathematical sense but information *about* things, interpreted in the context of stored knowledge, with potential consequences for behavior. The organism is not mechanically driven by stimuli from outside; it is interpreting these signals in its capacity as a self. The organism is meeting the world halfway, as an active partner in a dance that lasts a lifetime.

Fourth, causation in living systems is extended in time. We cannot build an explanation of what an organism does from an ahistoric description of its neural mechanisms. It is the way it is because of all the interactions that its ancestors had and that it has had with things in its environment. Through feedback from natural selection and through individual learning, organisms come to embody in their own physical structures knowledge about regular causal relations in the world. They act as causal capacitors, accreting causal potential that they can then deploy in the world. An instantaneous mechanistic view misses the central point of life—that is, that organisms do things for reasons, which arise from the past and are directed toward the future.

Finally, those reasons inhere *at the level of the whole organism*. We discussed in chapter three how a multicellular organism becomes the locus of fitness, from an evolutionary perspective. Natural selection doesn't care what goes on in individual cells or organs or systems, except insofar as they have an impact on the whole organism's survival and reproductive success. The meaning that the nervous system cares about is similarly ultimately grounded in those organism-level parameters. Of course, there are functionalities of different substructures, but they are judged by and shaped by their contributions to the goals of the whole organism. The reasons that emerge are thus the reasons *of the organism*, not of its parts. And the causal power that comes from having these reasons—from having paid attention to the causal regularities of the world—is similarly vested at the level of the organism itself.

## The Remaining Problem of Biological Determinism

What I presented to this point is a framework for thinking about agency in a naturalized way, consistent with what we know of biology, physics, and neuroscience. Though it is not often described as such, I claim that agency—the capacity of organisms to act with causal power in the world, for their own reasons—is the defining feature of life itself. It is, moreover, the bedrock on which we can build an understanding of free will in humans. For us to have the capacity to choose our actions absolutely requires that choosing actions be a possibility in the world in the first place. I hope the evidence and arguments presented so far have established that it is.

However, questions of free will in humans go well beyond the biology of agency. In particular, the concerns of biological determinism have not yet been addressed. We may be able to do things for our own reasons, but are we able to choose those reasons? Are we ourselves the cause of them? If our criteria for acting one way or another in any given situation are instantiated in the physical configuration of the brain, and if that configuration arises due to the combined cumulative influences of natural selection and our particular innate endowments, and feedback from our individual experiences, then are we really fully free in the present moment? Or are we in fact constrained by those prior causes that we did not control and we cannot now change?

If the configuration of our brains is determining what we're doing, then perhaps *we*—our conscious selves—are not really in charge. We will see in the chapters to come that this concern rests on a seductive dualist intuition: the idea that *you* are somehow distinct from the workings of your brain. A shift of perspective goes some way to alleviating this concern by presenting a more naturalized concept of the self. However, this reframing alone is unlikely to provide a satisfying justification for what most people mean when they use the term "free will." What people are usually after is some means by which we can really be in charge of our decisions on a moment-by-moment basis and not merely driven by our biology and the history that has shaped it.

I argue in the final chapters that evolution has provided exactly such a mechanism—or a suite of mechanisms—that grants us that capacity. We are not absolutely free, nor would we want to be—this is not a coherent notion at all, in fact. But we do have the capacity for reflective cognition, which means our subconscious psychology is not always opaque or cryptic to us. We have powers of introspection and imagination and metacognition that let us identify and think about our own beliefs and drives and motivations, examine our own character, and consciously adopt new goals or set new policies that guide our future behavior. We have, in short, the capacity of self-awareness.

And we have the capacity of self-control. We have, in real time, the means to intentionally adjust our behavior by selecting the objects of our attention and the different options for action that we consider and prioritize. None of this requires magic, any more than the biology of more fundamental forms of agency does. But it does rely on a recursive hierarchy of neural systems that has been most highly elaborated in humans, giving us a special place in the natural world.

# 10

# Becoming Ourselves

The preceding chapters present a picture of organisms as agents that act for their own reasons. That is, they can, at any given moment, do what they want. But can they want what they want? Do they really *decide* what they want to do? Or are they, in fact, just acting out the criteria embodied in the physical configuration of their brains at any moment? In relation to free will in humans, these questions were most forcefully articulated by Arthur Schopenhauer in his 1839 essay *On the Freedom of the Will* and were recently reiterated by erstwhile neuroscientist Sam Harris and many others, as described in chapter one. If we are really at the mercy of a long string of prior causes that have configured our brains in a certain way, then how free can we be?

The argument goes that if we were not ourselves in control of those prior causes—which include our species' evolutionary history and our own genetics, along with the effects of our upbringing and environment and experiences—then we did not freely choose our current inclinations and desires; that is, what we want. Rather, we are constrained to act in certain ways by the shackles of our histories, conditioned to act out the next scene in the story, and not truly free to author our own futures.

This line of thinking often leads to the existential fear that we—our conscious selves—are merely passengers along for the ride, with our brains in the driver's seat. Harris, for example, argues that we are merely witnesses to our thoughts and desires as they bubble up to consciousness, with no real access to the processes that determine them and no control over them. According to this view, we are just puppets at the mercy of our biology. And even if it is truly *our* biology—that is, it reflects purely internal constraints—this still limits or, for Harris, completely undermines the concept of free choice in the present moment. We will see later that this framing misconstrues the nature of the self. Indeed, if the kind of absolute freedom that Harris is seeking truly existed, it could not belong to anything we would recognize as a self.

In this chapter and the next I argue that we are not completely at the mercy of outside forces that determine our behavioral tendencies, nor are we completely driven by such tendencies at any moment. We ourselves, as agents, play an active part in shaping our own character over the course of our lifetimes. Moreover, we do have access to and awareness of many of our own motivations and desires. And we can develop and exert a faculty of self-control that puts us—a properly conceived notion of our *selves*—in the driver's seat.

## Is Who We Are "in Our DNA"?

The concern that we are programmed to act in certain ways by our biology has been reinforced by headline findings from behavioral genetics that consistently demonstrate the "heritability" of our personality traits. The (unintended) take-home message from such studies seems to be that our genes determine our psychology and our psychology determines our behavior. In fact, that framing overstates the case on both counts: our genes do *influence* our psychological predispositions and those predispositions do *influence* our behavior, but not in any direct or proximal way and certainly not deterministically. In addition, the focus on heritability strangely overlooks other factors that contribute importantly to our innate natures.

Heritability is a technical term that refers to the proportion of variation in some trait across a population that can be attributed to genetic variation. If we take adult height, for example, we can see that this trait varies across the population. Why is that? Well, we might also notice that tall people tend to have tall children and short people tend to have short children, immediately suggesting a genetic explanation. However, it is also true that rich people tend to have rich children and poor people tend to have poor children, and yet we ascribe that correlation mainly to environmental factors, rather than genetics. Twin and family study designs were devised specifically to try and tease out the contributions of genetic and environmental factors to variation in diverse traits.

They do so, for example, by comparing trait values between people who share the same family environment but who have differing degrees of genetic relatedness, such as pairs of identical versus pairs of fraternal twins. The consistent finding for a trait like height is that identical twins (who share all their DNA in common) are much more similar to each other than fraternal twins (who have co-inherited only 50 percent of their DNA, like normal siblings). Adoption studies use the opposite design, comparing many pairs of people with the same degree of genetic relatedness who either shared or did not share the same family environment. Again, for height, the finding is that individuals strongly resemble their biological relatives and share almost no excess similarity with their adoptive relatives.

From data like these and, more recently, from studies across large populations of only very, very distantly related people, it is possible to estimate how much of the variance in a given trait is due to genetic variation in a given population (i.e., the heritability), how much is due to variation in family environments, and how much remains unexplained by either of those factors. Unsurprisingly, the heritability of height is about 80 percent—that is, the vast majority of the variance we see in this trait across people is caused by differences in their genes. Only a very small amount of variance is due to variation in the family environment, at least in the population samples that are studied. Note that this does not mean environmental factors are not important—it's just that they don't vary much within the population samples in those kinds of studies. Environmental factors certainly do influence the average value, however; indeed, variation in height among different countries or among populations over time correlates strongly with environmental factors, such as protein intake during childhood.

This highlights an important fact: the heritability of a trait is not some kind of biological constant. It applies only to the population studied and can vary under different conditions. Moreover, the finding that much of the variation in a trait in a given population is due to genetic variation between individuals also does not imply that any differences in the *average value* of that trait *between populations* are due to some consistent set of genetic differences between those populations. And finally, heritability does not apply to individuals—it is not that 80 percent of your height comes from your genes and the rest from your environment; that is a nonsensical statement. It applies only to the sources of variation in some trait *across a population*.

The same kinds of studies have been done for all kinds of psychological or behavioral traits. They require in the first instance that researchers define and measure something that qualifies as such a trait. It might be a score on a test of some sort, such as an intelligence quotient (IQ) test. Or perhaps it is some kind of statistical construct derived from responses to a series of questions about a person's behavior, which are designed to tap into personality traits like extraversion or neuroticism (there is much more later on what such constructs might reflect). Or it could be reports of actual behaviors or events or *life outcomes*, such as whether a person is married or divorced, or arrested or incarcerated, or how many years of education they've completed, or what is their annual salary.

As I described at length in my previous book, *Innate*, the results of genetic studies for all these kinds of measures—cognitive performance, personality traits, or diverse life outcomes that reflect in some way a person's own behavior—are remarkably consistent. They show a general pattern of moderate heritability, with anywhere between 30 to 60 percent of the variance in the measured trait attributable to genetic variation in the population. They consistently show very little effect of the family environment, which usually explains only 0 to 10 percent of the variance. Notably, this leaves around half of the variance unexplained by either genetics or family environment.

What are we to make of these findings? Do they support the kind of genetic fatalism sometimes adopted in relation to free will? Often they are presented as a radical revelation that should change the way we think about our psychology and our behavior. Conversely, much ink has been spilled arguing that the methods used in behavioral genetics studies are flawed or biased in one way or another and that the demonstration of heritable variation in our psychological traits is not trustworthy. But really such variation is entirely to be expected—indeed, it is inevitable. If we accept that human nature—that is, what sets us apart, biologically, from other species—generally is genetically encoded, then we must expect some innate variation on that theme across individuals.

Evolution has shaped the genomes of all species of animals, such that a fertilized egg with a given species' DNA will develop into an organism of that species type with its characteristic physical form. That includes the physical form of the brain, with all its intricate species-specific wiring that confers the behavioral capabilities and tendencies of each species. Frog DNA directs the formation of a frog with frog nature. Scorpion DNA programs scorpion nature. And human DNA programs human nature. This is not to say that environment and culture do not allow us to build on that nature and affect how it is expressed—they certainly do. The point is just that we are endowed genetically with human-typical behavioral capabilities and tendencies (which include the capability and tendency to learn from our experiences) by virtue of having human-typical brains, the self-assembly of which is directed by our human-typical genomes.

But we don't all have identical genomes. Genetic variation arises in each generation due to replication errors that occur when DNA is copied as eggs and sperm are generated. Each new individual thus carries a set of brand-new mutations from his or her parents, as well as the ones they carried from their parents, and their grandparents, and so on. You might wonder then how the information in the genome doesn't simply degrade to nothing over time. The answer is that the accumulation of genetic variation is counterbalanced by natural selection. New mutations that have very deleterious effects—that seriously impair the program of development or the physiology of cells or tissues—are rapidly selected against. Individuals who inherit such mutations either die or are so severely affected that they tend not to reproduce, and thus the mutations are not passed on.

At the other end of the spectrum of severity are changes to letters of the DNA that have no effect at all. Only about 3 percent of the human genome actually codes for the proteins that carry out cellular functions, including the instructions for how and where each protein is expressed. In those regions, the precise sequence of DNA bases may be very important. But in much of the genome, changing a base here or there doesn't matter at all. New mutations in these regions may spread through the population effectively at chance.

It is variation in the middle of that spectrum of effects that makes the biggest contribution to our traits, including our psychological traits. These latter traits are generally affected by variation not in one or two specific genes but in many thousands of genes, all at once. Each of these genetic changes or variants may have only a tiny effect by itself, which explains how they can persist in the population. If new mutations have any effect it all, it is much more likely to be deleterious than beneficial: it's hard after all to improve by random tinkering on billions of years of evolution. But if the effect is small, then natural selection may not get much purchase on it, given the background of other variants and random variation in life success. Such variants can persist and spread to a certain level in the population, and although their individual effects are small, their collective effects can be quite large.

The result is that most human-species-typical traits inevitably show variation around a mean. For example, humans tend to be human sized, not mouse sized or elephant sized, but we are not all the same height. We have limbs of a certain average length and noses of a certain average shape and so on, but those parameters all vary around those averages. There is no way they could not differ. Natural selection cannot keep all mutations out of the genome—if it did, there would either be no one left to breed, or if it had somehow been able to select for perfect, error-free processes of DNA replication, there would have been no raw material for evolution to work with in the first place.

This same type of variation is seen in our brains. The canonical "human genome" encodes a program to make a canonical "human brain." But no individual actually possesses such a genome or such a brain. We all have our own version of "the" human genome and consequent variations in the way our brains are put together. The heritability of psychological traits is simply the manifestation of that fact: people who have more similar versions of the genome (who are related to each other) have brains that are demonstrably structurally *and functionally* more similar to each other. This should be no more surprising or controversial than the fact that people who are related to each other have more similar faces.

## Heritability Is Not the Whole Story

However, focusing on heritability in the conversation about free will misses several important points. First, genetic variation is not the only source of innate neural and psychological differences between people. Even people with the exact same genome—that is, identical twins—do not emerge with identical brains. The program of development that directs the self-assembly of the brain is staggeringly complex, involving the actions of thousands of proteins mediating the patterning of the brain into different regions, the differentiation of thousands of different cell types, the migration and spatial organization of these cells, the extension of axons and dendrites, and the eventual formation of synaptic connections, as well as controlling the biochemical processes of plasticity that refine that initial wiring pattern. All those processes are subject to some level of random noise at the cellular level, as proteins diffuse around and bind and unbind to other proteins or to the DNA to regulate gene expression, and all manner of other biochemical interactions and chemical reactions take place, effectively in a probabilistic fashion, set by the relative concentrations and affinities of the various elements.

The genome has no control over those processes. It cannot encode a definitive outcome, only the biochemical processes by which self-assembly takes place. There isn't, in fact, anywhere near enough information in the DNA sequence to specifically encode the individual connectivity patterns of all the cells in the brain. The best that natural selection can do is ensure that the developmental program is robust enough to produce a generally viable outcome most of the time. Indeed, it has done exactly that: most of that cellular noise is effectively buffered to keep the outcome within an acceptable range. But the noise does collectively produce *variation within that range,* even when development

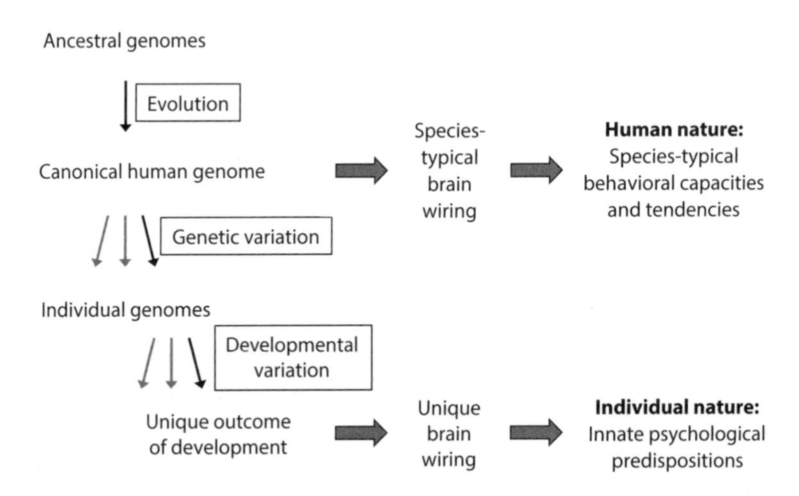


FIGURE 10.1. Innate predispositions. Human nature generally—the set of behavioral tendencies and capacities that characterize us as a species—has been shaped by millions of years of evolution. This is realized through a genomically encoded program of development that leads to a human-typical pattern of brain wiring. However, each of us develops a unique version of this pattern, thanks to both genetic and developmental variation, endowing us with an innate set of psychological predispositions—our individual natures.

starts from the exact same genome, as in cloned animals or identical twins in humans.

Most of the resulting variation in the brain is quantitative in nature—some area of the brain might be a little bigger, or some regions might be connected by more axons in one twin than the other, for example. But because brain development is a very nonlinear process, with many processes contingent on prior events having happened a certain way, small variation in some parameters during development can sometimes be amplified and lead to qualitatively distinct outcomes that may be reflected in behavioral traits (see Figure 10.1). One twin might end up left-handed, for example, while the other is right-handed.

The upshot, as described in detail in *Innate*, is that developmental variation is an important additional source of variation in the way our brains get wired and in our resulting innate psychological predispositions. In twin and family studies, this kind of random developmental variation may account for much of the variance that is unexplained by either genetics or family environment. For the purposes of our discussion of free will, however, this may seem to create an even larger problem than the demonstration of (the only partial) heritability of psychological traits. We really are not born as blank slates, both because of inevitable genetic variation in the program of brain development and the equally inevitable variation in how that program plays out.

This seems only to reinforce the notion that we are constrained to act in specific ways by factors over which we had no control. I argue later that this is an overly simplistic view of how our innate psychological predispositions actually influence our behavior. But the primary concern driving this notion also seems misplaced in two ways.

First, the variation that we see in our individual natures is tiny relative to the similarities in human nature generally. Of course, we tend to focus on that variation because we naturally compare ourselves to other humans. We may say, "I wish I were more confident like Gary," or "I wish I were more conscientious like Mary." We may thus see our own personality traits as constraints relative to how other people behave. But we ignore the much more massive set of constraints that come with simply being human. We rarely think, "I wish I were more sociable like a naked mole rat," or "I wish I were more chilled out like a walrus." If we're going to worry about innate constraints on the ways we behave, it seems odd to focus on just the relatively minor variation between humans when evolution has carefully sculpted all our natures generally, just as much as any other species. There seems to be little concern that we are constrained to behave like generic human beings, but strangely more worry that we are constrained to behave like ourselves.

Second, the now-defunct alternative view—that we really are born as blank slates—is usually accompanied by the idea that *our personalities are shaped instead* by our upbringing and formative influences during our early lives: by nurture. This typically involves the idea of critical periods during which our personalities are molded, beyond which our patterns of behavior become entrenched and further change becomes difficult or impossible. Thus, in this alternate model, we still do have personality traits that still do constrain our behavior along typical lines, and they still are caused by forces and factors outside our control. The outlook for unconstrained action thus does not seem much rosier under this view.

Heritability is thus something of a red herring. If there is a challenge to our notion of free will that arises from prior causes shaping our psychology, it stems from the much more general and more sweeping constraints that come with being human, and it holds whether the variations that define our individual personalities are determined by nature or nurture. Indeed, the framing of nature and nurture as opposing forces is extremely misleading and unhelpful. The truth is that our innate predispositions and our experiences interact in an ongoing fashion over our lifetimes in ways that cannot be untangled.

Our innate psychological predispositions do not determine our behavior on a moment-to-moment basis. What they do is influence how we interact with the world and adapt our behavior to it over the course of our lifetimes. They shape how we become ourselves—the way our character and our habits emerge—in interaction with our experiences and the environments we encounter, choose, and create for ourselves. We are not passive in this process, merely acted on by causes outside our control: this process is one in which we ourselves, as active agents, play a causally effective role.

## Personality Traits

To understand the dynamics of how our character emerges, we should look first in more detail at the underlying personality traits most commonly defined by psychologists. For millennia, people have sought to classify the architecture of human personality. Our common experience is that individual people have reasonably stable, characteristic patterns of behavior: broad tendencies and propensities to act in certain ways across diverse situations. To account for the diversity of these patterns, it is natural to think that people may vary along a finite number of independent dimensions.

The ancient Greek physician Hippocrates (followed by Galen) defined four *humors* that vary in levels among people and determine their associated temperaments: sanguine, choleric, melancholic, and phlegmatic.

Somewhat more scientific attempts at classification were also made throughout the centuries; for example, using statistical analyses of the more than eight thousand words that refer to personality traits in the English language to see how many clusters they fall into. For example, the adjectives "bubbly," "lively," "vivacious," "outgoing," and "excitable" all seem to refer to overlapping, if not identical, traits. The definition of such clusters, along with the development and statistical analysis of questionnaires that probe various aspects of people's behavior, has led to a number of schemes of personality in modern psychology. These schemes vary in how many major dimensions they identify, with the most popular known as the "Big Five."

These five dimensions are called Extraversion, Neuroticism, Conscientiousness, Agreeableness, and Openness to Experience. It's important to emphasize that this classification is not written in stone, nor is it universally agreed on: different schemes identify anywhere from two to sixteen supposedly independent major dimensions. But the Big Five will do for our purposes here. Each of these traits reflects many subfacets of behavior, with the idea being that correlations across these subfacets reflect, in part, a singular underlying latent factor that influences them all. For example, questions relevant to extraversion may not only ask how much you like to socialize (the most obvious question, given the colloquial understanding of the term) but also how much you enjoy sex or how interested you are in travel, or they may probe how talkative or active or assertive you are. These are all distinct and partly independent facets (you can be assertive without being talkative or sociable, for example), but across large samples of people, positive answers on one facet tend to be correlated with positive answers on others within this group.

A higher level of neuroticism reflects a greater sensitivity to negative emotions, tendency to worry, emotional volatility, and vulnerability to stress. Higher conscientiousness reflects greater focus, self-control, and a desire for order and predictability. Higher agreeableness reflects a more cooperative nature; a desire to be helpful, kind, and considerate; and a general concern for social harmony. And higher openness to experience reflects a more aesthetic sensibility and greater imagination, curiosity, and a willingness to try new experiences.

By summing up people's responses to different questions and the correlations among them, psychologists can infer the strength of the supposed underlying latent factor and assign an arbitrary numerical score for each defined trait. These scores vary, by convention, along a single linear scale, with the values across the population showing a *normal distribution* or bell curve: most people are near the middle, and fewer individuals score very high or very low. These measures are fairly reliable, correlating well if an individual is queried multiple times, and they are also fairly stable over time in individuals. Although some broad dynamic trends with age are seen across the population, such as an increase in conscientiousness and a decrease in neuroticism over people's lifetimes, the relative rank of individuals remains reasonably stable. And scores correlate well whether people rate themselves or are rated by someone who knows them well.

Thus, these constructs seem to have enough validity to be scientifically useful; for example, in looking at how similar relatives are to each other in their personalities. As described earlier, these measures are moderately heritable, with genetics explaining somewhere under half of the variance across the population. This really isn't surprising: there is a reason why we have common phrases like "the apple doesn't fall far from the tree" or "he comes by it honestly" or "it's not from the wind she took it." Everyday observations of family resemblances foster a fairly accurate folk understanding that personality has a genetic component to it.

More surprising, at first blush, is the equally consistent finding from twin and family studies that the family environment has very little effect on these traits. One might have expected our upbringing to also have some influence on these personality measures, but that does not seem to be the case. For example, adoptive siblings are hardly correlated at all with each other on these traits, and identical twins who are reared apart tend to be about as similar to each other as ones who were reared together. This finding has been interpreted as implying that "parents don't matter" or don't have any effect on their children's psychology and behavior. This is a drastic overextrapolation, in my opinion. We will see later that upbringing can affect many aspects of the way we behave without altering those underlying traits themselves.

Indeed, the whole point of these constructs is that they are supposed to tap into latent, underlying biological parameters that vary in a stable manner across individuals and that manifest in characteristic behavioral tendencies across very diverse situations. By design, these constructs abstract these patterns away from the particularities of each individual's life. It is thus not that surprising that the measures are not very responsive to differences in upbringing or other aspects of the family environment. They also do not appear to be reliably correlated with any systematic factors in the wider environment, which suggests they really are inherently stable and refractory to change. Consistent with this view, I argue that much of the remaining variance that is not explained by either genetics or family environment is due to inherent developmental variation, thus making these traits even more innate than genetics alone would suggest.

The question remains: What do these Big Five traits actually represent, biologically speaking? And how do they relate to the mechanisms we explored in previous chapters that allow agents to direct their own behavior, make decisions, and select actions? There are two schools of thought on what these statistical measures reflect. The first is, as I outlined earlier, that they reveal a singular underlying latent factor—a real biological thing—that influences many different behaviors. The second inverts this relationship, suggesting that these measures are just statistical constructs that capture variation in many underlying biological parameters that collectively influence broad domains of behavior. For me, the evidence favors the second view.

If traits like extraversion and neuroticism really reflected some singular factor in the brain (such as "positive emotionality" or "negative emotionality," for instance), one might expect to find some brain measure that consistently correlates with them—say, a bigger amygdala or higher levels of serotonin or more dopaminergic connections with the striatum, or really an infinite range of other possibilities. Researchers have looked for many years for these kinds of correlates and effectively come up empty. There are reports of positive correlations but none has been robustly replicated. A reasonable conclusion, at some point, is that a trait like extraversion is not in fact a singular "thing in the brain." Rather, it is a construct reflecting multiple neural parameters.

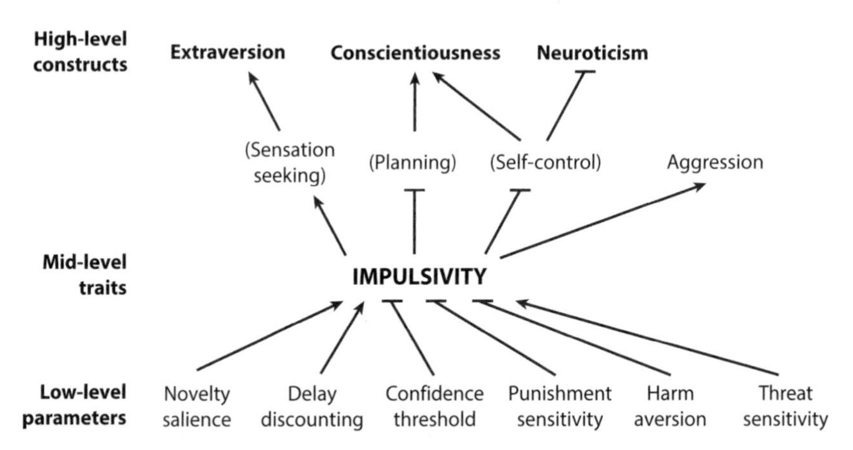


FIGURE 10.2. Impulsivity. This hypothetical model shows impulsivity as an arbitrarily chosen mid-level trait that is affected (positively, arrows; or negatively, T-bars) by the levels of many lower-level decision-making parameters and that, in turn, feeds into higher-level personality constructs.

We already encountered in our discussion of the circuits controlling decision making and action selection what some of these parameters could be. An animal navigating through its environment has to monitor its own internal state to generate motivations; assess opportunities and threats; anticipate rewards and punishments; weigh multiple, often conflicting short- and long-term goals against each other; assess the reliability of different kinds of perceptual evidence; attach some measure of certainty to its various beliefs; weigh the confidence level required to act versus the urgency for action; consider how much effort it is willing to expend or how long it is willing to wait or how much risk it is willing to accept for a given reward of a given probability; and so on (see Figure 10.2).

As we saw, these operations are implemented through the activity of diverse, interconnected neural circuits, and, like all other aspects of the brain, these circuits inevitably vary among individuals due to both genetic and developmental variation. That means that the tunings of factors like reward sensitivity, threat sensitivity, novelty salience, confidence thresholds, delay discounting (how long you are willing to wait for a reward), and risk aversion naturally—innately—vary among people. Those kinds of parameters are things we do "find in the brain," in the sense that neuroscientists have identified at least some of the circuits that seem to mediate these different signals and feed into the processes of belief formation, goal prioritization, and action selection. Indeed, new technologies in animals are allowing researchers to observe these separable functions and to correlate levels of activity in specific circuits with, say, confidence levels, assessed threat levels, or anticipated reward; scientists are even able to experimentally tweak the tuning of these parameters to alter decision making and behavior in real time.

Many of the relevant signals are carried by neuromodulatory circuits that use chemicals like dopamine or serotonin or norepinephrine as signals. There is a general sense—in my view promoted by the way these kinds of "brain chemicals" are discussed in the popular press—that these neuromodulators act like hormones inside the brain. That is, they are released in response to some stimuli and simply bathe their target areas (or even the whole brain), indiscriminately affecting all the neurons in that region. This may be why people often talk of *levels* of dopamine or serotonin, as if they are a unitary signal with only one function or meaning (such as reward in the case of dopamine).

There is some truth to that conception: these neuromodulators do act sometimes by *volume transmission* and can have widespread effects. But we are increasingly learning that many subcircuits of these neuromodulatory neurons have very precise connections, their release is tightly controlled at specific synapses, and the effects of the signal are determined by the specific repertoire of receptors that each neuron expresses. The meaning of these signals varies across these subcircuits. For example, dopamine release in the anterior and ventral portions of the striatum does signal reward (or predicted reward or the deviation from a predicted reward, depending on your model), but release in more posterior regions actually signals threats. Of course, how those signals are interpreted and acted on depends on the organization of the receiving circuitry and interconnected systems. All of which is to say that there is ample opportunity for these modulatory systems and the decision-making circuits they feed into to vary in complex and subtle ways across individuals.

The profile of variation across all these parameters, under this model, feeds into an overall level of extraversion or neuroticism or conscientiousness. Two people might thus end up with similar scores on any of these traits for quite different underlying reasons. Because variation in all the individual parameters is heritable, the higher-level constructs are too.

So, yes, we really are all tuned a little differently. And these differences affect our behavioral tendencies in any given situation. But here's the thing: we are never actually in "any given situation"—we're always in *some particular situation*. That's what life is, after all, one damn thing after another. What we actually do in any particular situation depends on all kinds of factors, including our evaluation of the current scenario in the context of the knowledge we built up from our own personal history, along with our current goals and motivations, the predicted utility of the outcomes of various actions, and so on. Within this complex context, our personality traits are a contributory factor, but they are far from a determining one in any moment. However, they do have an influence on the development of our characteristic adaptations and habits over time.

## Habits, Heuristics, and Policies

Habits get a bad rap. Most of the time our discussion of them is in relation to *bad habits*—things we do that we know are bad for us or cause harm or hurt to other people and that we wish we could override. Examples are smoking or overeating or procrastinating or any of a wide range of behavioral ruts that we can't seem to get out of. We do not tend to notice or talk about our good habits: all the things we do that are simply useful, efficient automations of routine tasks that leave our cognitive resources free for handling new situations.

As discussed in chapter six, most of our behavior is habitual in nature. Reinforcement learning mechanisms evaluate the outcome of an action and increase or decrease the likelihood that you will repeat the same action the next time you are in the same or a similar situation. If a given action continues to have a positive outcome, then that reinforcement will eventually lead to it becoming habitual. This means that the processes of decision making and action selection are preempted: there's no need to expend energy and time and limited cognitive resources deciding what to do, weighing up a load of possible actions and their likely outcomes, when you already have a perfectly good option primed and ready to go.

One way that scientists define a habit is as a behavior that is simply cued by some kind of stimulus or environment or situation. Studies of animals often involve reinforcing some behavior (typically a very simple action) many times in association with some cue so that, after some time, when the animal perceives the cue it directly and rapidly triggers that action. Studies of the neural basis of this process, in humans and other animals, reveal that it often involves a shift from what is known as *model-based reasoning*, where individuals use their model of the world to figure out what they should do, to *model-free reasoning* in which the model is not consulted or used to inform action and, instead, a habitual response occurs, like a reflex. This kind of experimental paradigm is highly controllable and reproducible and thus provides an excellent platform for investigating underlying mechanisms.

However, there are other kinds of behaviors that can still be called habitual but that do not involve the direct triggering of a very specific action by some very specific cue. Instead, they involve a somewhat more abstracted kind of recognition of a general *type of situation*, which may prime a general *type of response*. For instance, when you wake up in the morning, you may be in the habit of having breakfast before going to work or school, but what you have for breakfast and where you have it may vary from day to day. It is not a specific motor action that is triggered but a habitual activity. I certainly don't need to think every morning about whether I should have breakfast—I've learned that it's a good thing to do or I'll be hungry and irritable in short order. I'm happy with that practice, and I can execute it with effectively no cognitive effort.

Most of our daily activities are like that—routine and habitual—even if the specific actions that constitute them can vary. The psychologist and economist Daniel Kahneman refers to this mode of thinking as *System 1* and characterizes it as fast, efficient, and requiring little conscious oversight. That's not to say that such habitual activities cannot be altered; if necessary, a more deliberative mode of thinking, termed

*System 2,* can be activated: it is more effortful, slower, and more conscious. For example, if you encounter a detour on your drive home, you can easily switch from automatically following your normal route to actively deliberating about alternatives.

Many of our longer-term patterns of behavior are similarly habitual, following general *heuristics*—rules of thumb about the best way to behave in various types of circumstances—or they are more long-term commitments and policies that were decided well in advance of the present action. For example, say Maya decided to go to university to become a doctor. She knows that will be a hard road that will require lots of work and dedication, and she's committed herself to that task, even though it will take years to complete. Her behavior on any given day is thus guided by (and constrained by) that prior commitment. She may have, moreover, decided that the best way to ensure success is to attend all her lectures, even the really early ones that getting up for is frankly torture. So, when she wakes up on a Tuesday morning, she doesn't have to decide what to do. Her goals are set: she wants to become a doctor, so she has to complete her university training, so she has to go to all her lectures, so she has to get up and catch the 7:30 A.M. bus.

Many of the decisions we make are similarly self-regulating: they are decisions *about our future decisions,* constraining our choices in future scenarios by directing them in a sustained way toward some end. If we didn't have this ability to constrain our own future behavior, we would not be capable of maintaining long-term goals. Some short-term imperative, like staying in our warm bed on a wintry morning, would always interrupt what Jean-Paul Sartre called our "projects." These encompass our goals, as well as the plans we make to pursue them, which direct the activities we prioritize, in turn constraining the specific actions we consider and choose at any moment.

These kinds of commitments are informed by our underlying psychological predispositions. In the process of adapting our behavior to our circumstances and navigating the opportunities available to us, we are naturally led by our own interests and aptitudes and preferences, taking a hand in creating circumstances that suit us, insofar as that is possible. For example, the ability and inclination to study for many years to become a doctor may reflect or even require a high level of the conscientiousness personality trait, reflecting any number of underlying facets such as the ability to delay gratification, persistence, attention to detail, low impulsivity, and low novelty seeking. But aptitude for that career will require more than just that innate predisposition: it will also require the *habit* of applying oneself in the academic sphere and the resultant skills to which that habit leads. It's easy to see how such a habit could emerge; someone with a high level of conscientiousness will naturally tend to work hard at school. Their parents may have similarly high levels of this trait and so recognize and reward the child's efforts. Thus encouraged, the child will continue to engage in this behavior, leading to further successes and establishing a reinforcing cycle.

Psychologists Dan McAdams and Jennifer Pals refer to these progressively emergent habits and patterns of behavior as our *characteristic adaptations*, reflecting the interplay between individuals and their worlds. There is no evidence or even suggestion that these adaptations change our underlying psychological traits; instead, they represent the process by which those traits are expressed in our actual behavior. We are active players in this process—our experiences are not just things that happen to us. They are things we engage in, and as we develop greater autonomy and self-directedness through our childhood and into our adult lives, they are things we actively choose. Similarly, our environments while we are young may be out of our control, but environments are not somehow assigned randomly to different people. First, because we tend to psychologically resemble our relatives, including our parents, even our earliest environments may reflect and reinforce our own predispositions, as in the earlier example. Second, as we mature, we select and even create our own environments according to our interests and inclinations within, of course, the constraints afforded by circumstances.

There is thus a coevolution of individuals with their environments, similar to the coevolution of species and their environments seen over evolutionary timescales. The upshot is that our patterns of behavior are not simply driven, in the present moment, by our innate predispositions. They emerge through time as we each, in our individual ways, simultaneously adapt to and shape our environments and circumstances as we develop habits of thought and habits of action—heuristics and policies and commitments—all of which inform and guide, but also necessarily constrain, our behavioral options. Beyond those habits of behavior, which are fitted to more or less specific circumstances, we also develop habits that apply much more broadly: these are habits of character.

## Character

The distinction between personality and character is a tricky one, especially because the two terms are used in overlapping fashion in common parlance. Here, I intend personality traits to refer to underlying psychological predispositions of the kind we have been discussing. These denote differences between individuals that are more or less neutral; that is, it is not necessarily *better* to be higher or lower in extraversion or openness to experience or even conscientiousness or neuroticism or agreeableness. There is not a particular level of each of these traits or a single combination of levels that is optimal in all scenarios; it may, for example, be quite adaptive to be high in neuroticism in circumstances where threats abound. Particular levels of these traits are thus not consistently *instrumentally* better or worse in terms of their usefulness for the individual, and they are also not seen as *morally* better or worse: it doesn't make you a better or worse person to be one way or another for traits like this.

Character traits, by contrast, are essentially defined as morally better or worse; in fact, they are typically categorized as virtues or vices. The virtues include things like honesty, fairness, courage, humility, generosity, steadfastness, loyalty, integrity, prudence, patience, forbearance, temperance, selflessness, and so on (and the vices their opposites). A judgment value is implicit in each of these terms: they are seen as good or bad. This needn't imply some absolute frame of moral reference, however, somehow just given by the universe or, as is often proposed, a divine Creator. These judgments arise in a much more pragmatic fashion, because we are an obligately social species.

Virtues are essentially prosocial traits: it's not that they are good or bad in an absolute sense or even that they're good or bad for the individual—it's that they're good or bad for everyone. Humans survived and thrived and colonized practically every niche on the planet by working together. That required cooperation over extended periods of time, concerted action, sharing of resources, division of labor, and so on. When everyone cooperates, everyone benefits. This tendency may have evolved from a natural instinct to look after our offspring and our relatives in extended family troops. Any genetic variants that predisposed to this kind of behavior would naturally be spread as they are shared between relatives. Over some period, that natural tendency to cooperate with kin may have extended to other, unrelated members of a group, driving social innovation and helping humans dominate the landscape.

The trouble is that it's quite easy in this kind of system to be a freeloader or a cheater and, with less effort, do a little bit better than everyone else. Natural selection will reward individuals who develop behavioral strategies that let them take advantage of the cooperativeness of their fellows to outcompete them. However, in a kind of arms race of justice, natural selection also seems to have led to the development of cognitive resources that could detect such cheaters: the ability to keep track of who's done what, who's given what to whom, and who's done their share of the work or shirked their responsibilities. Humans—even toddlers—and some other animals seem to have an innate sense of fairness, coupled to an outrage engine that gets engaged when cheaters are detected. As argued by philosopher Patricia Churchland, the formalized systems of morality seen across human societies may have evolved from these basic prosocial tendencies and capacities and the need to monitor and enforce prosocial norms.

In addition to pro- or antisocial aspects, the other main theme among character traits is one of self-control, as exemplified by patience, prudence, temperance, foresight, and perseverance. These traits represent the triumph of rationality over more basic drives (a capacity that may underpin more overtly prosocial behaviors). Rationality has been recognized from the time of Plato and Aristotle as a defining capacity of humans, elevating us above the beasts. Psychologist Albert Bandura describes these self-regulatory aspects of character as he explains, "An agent has to be not only a planner and forethinker, but a motivator and self-regulator as well. Having adopted an intention and an action plan, one cannot simply sit back and wait for the appropriate performances to appear. Agency thus involves not only the deliberative ability to make choices and action plans, but the ability to give shape to appropriate courses of action and to motivate and regulate their execution."[1]

This capacity needs to be actively developed. Indeed, a defining aspect of all character traits is that they do not solely reflect innate endowments. Individual differences in personality may certainly have an influence on them, but they represent mature habits and skills and practices more than just predispositions. The (perhaps apocryphal) ancient Taoist figure Lao Tzu is quoted as saying, "Watch your actions; they become habit. Watch your habits; they become character. Watch your character; it becomes your destiny."

Questions of moral character and how it should be developed feature prominently in most religions and schools of philosophical thought. It is not that any one of them has a claim to truth in its specific views on how people should behave. The point is just that it is widely accepted that character traits are not just given but need to be actively developed and, moreover, that *individuals themselves* have both the power and the moral responsibility to engage in and direct that development.

The ancient Roman philosopher and statesman, Cicero, dwelt extensively on this topic, especially in his last book, *De Officiis* (On Duties), which he wrote to his son Marcus, then a student of philosophy in Greece. He highlighted the cardinal virtues of justice, temperance, wisdom, and fortitude as ones that should be aspired to, naturally situating examples of relevant behaviors in the context of Roman society. For Cicero, character emerged from four sources, which should be recognizable from the preceding discussion: human nature generally (including the capacity for rational thought and action); our own individual natures; events, experience, and the environment; and finally, the *accumulating effects of our own choices.*

1. Albert Bandura, Social cognitive theory: An agentic perspective. *Annual Review of Psychology* 52 (2001): 8.

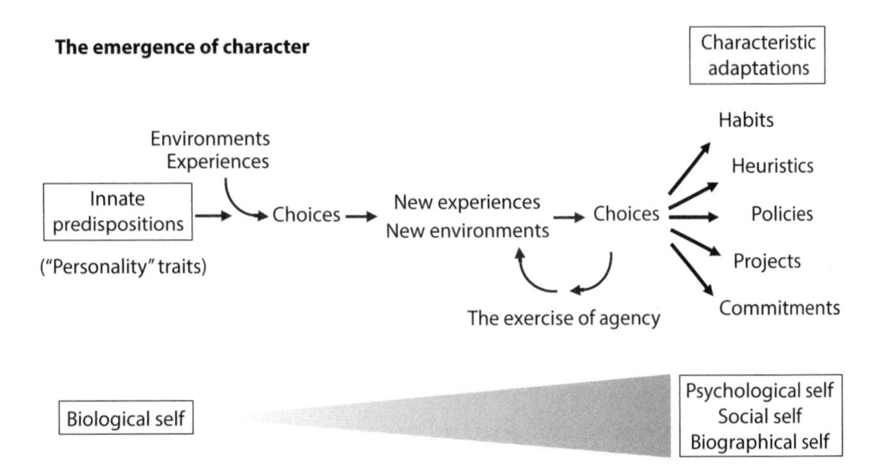


FIGURE 10.3. The emergence of character. Our innate predispositions influence our actions, experiences, and responses to those experiences. The interplay between these factors, driven by the ongoing exercise of our own agency and expressed through the idiosyncracies of each of our lives, shapes the emergence of our characteristic adaptations: the habits, heuristics, policies, projects, and commitments that define who we are as persons and that contribute to the continuity of our behavior as selves, through time.

Of course, this kind of character development is often informed by explicit moral instruction, shaped by the specific society and culture that an individual grows up in (see Figure 10.3). More generally, it is reinforced by constant feedback from external sources. Much of the debate on free will centers on questions of moral responsibility—whether people really deserve praise or blame for their actions. We will return to this question in the final chapter. But what is interesting to note here is simply that praise and blame exist: they are things we engage in, all the time, in fact. We're constantly judging whether people should or shouldn't have done something and are often willing to expend some effort in letting them know about it. If personality were simply fixed or given, there would be no point in praising or blaming people because that would have no effect on their behavioral tendencies. But if character emerges over time, through individuals interacting with the environment, then these kinds of reinforcement signals will be key in allowing individuals to evaluate their own behavior with reference to social norms.

These reinforcement signals are particularly relevant for children as they develop their habits of character. Through our approval or disapproval, we teach children that some activities are good or bad, reinforcing prosocial norms and self-regulatory behavior: being considerate, sharing, taking turns, being patient, thinking about the consequences of your actions before you act, and so on. Approval or disapproval tends to engender feelings of pride or guilt and shame—the so-called self-directed moral emotions. Along with more general emotions associated with evaluating outcomes such as regret, satisfaction, disappointment, and frustration, these act as strong learning signals for future behavior, which is guided by the experience-based *anticipation* of these emotions when considering various options for action. The external signals from other people's reactions thus get internalized to form part of a self-regulatory apparatus—a moral compass—that we use to guide our own behavior.

Overall then, the idea that our innate personality traits *determine* our behavior is frankly simplistic. In essence, this view sees people as passively driven by their biological predispositions, on a moment-to-moment basis, like robots that are tuned one way or another. The character viewpoint, by contrast, puts people *as agents* at the center of a process of engagement and coevolution with the world, actively developing the habits, attitudes, policies, and mature tendencies that make us all who we are. These views can be reconciled by recognizing that there will be an interplay between our innate traits and the trajectory of development of our character. Cicero was at pains to say that the choices each of us makes for how we are going to live our lives should align with our individual natures: "Everyone has the obligation to ponder well his [*sic*] own specific traits of character. He must also regulate them adequately and not wonder whether someone else's traits might suit him better. The more definitely his own a man's character is, the better it fits him."

The idea that our current psychology was shaped by prior causes *over which we had no control* is thus not accurate. Each of us has been very actively involved in the process of shaping the person we've become.

## Being a Self

The arguments in this chapter suggest that we do have a hand in shaping our own character; not all the prior causes are things that were out of our own control. However, it could still be argued that the presence of *any constraints,* right now, regardless of where they came from, invalidates the idea of *real* free will. This idea hinges on an absolutist notion: we are only truly free if we are completely free from any prior constraints. Then, we not only can choose what we do based on what we want, but also can freely choose *what we want,* unfettered by . . . well, anything.

Once you dig into this notion a bit, it quickly becomes incoherent, revealing a dualist notion of the self that evaporates when examined closely. It implicitly identifies the self with the conscious mind, construing all the other machinery of agency that we have been discussing as non-self—just physical matter—really a cage restricting our freedom, as opposed to the vehicle of it.

Even if your conscious self were somehow able to choose *what to want to do,* free from any constraints arising from prior causes or subconscious influences, then on what basis would you decide? If you're not constrained by your own character, or informed by your past experiences, or committed to any long-term goals or policies, then how are you deciding? On a whim? Or based on something?

Presumably you'd like to be able to decide for some reasons, for *your* reasons to do whatever you feel like, but why would you feel like anything? Maybe you'd like to exercise some preferences about what you'd like your reasons to be. But then we're back where we started. Conscious or not, reasons (or preferences) are constraints—that's how they guide action, making some choices more likely than others. Totally unconstrained action would be totally uninformed action. It's not clear it would even qualify as action, in fact. An action is something an agent—a self with aims—does. Otherwise it's just a physical system behaving effectively at random. That doesn't really sound like you deciding; it's not obvious where *you* are in that picture at all.

There seems to be a desire for the self, in any given moment, to be free in its choice of action from all prior causes. You might like to be released from all the commitments your past self has made, from the habits you've forged, from the underlying traits that inform your character. And you might prefer to not be beholden to your future self by the decisions you make in the present. But once those links are broken, you would find yourself entirely adrift (see Figure 10.4). There is no self in a moment. The self is defined by continuity through time. You, right now, in the present, are just the momentary avatar—the representative in the world—of a self that stretches from the past to the future.

Continuity is the defining property of life. In a unicellular organism, the whole imposes constraints on the parts: all the interlocking feedback interactions keep all the biochemical processes organized in a certain pattern. The organism is not a pattern of stuff; it is a pattern of interacting processes, and the self is *that pattern persisting*. The same is true at a higher level—not just physically but also psychologically. What else does it mean to be you other than to think like you and behave like you in some consistent manner through time? Who would you be if you did not? This continuity of character *defines you*—it defines each of us, so much so that when people act out of character, it is often a sign of some pathology like schizophrenia or dementia or a brain tumor (or, for fictional characters, a sign of being badly written).

That doesn't mean that we don't change and grow over time—of course we do. But we do so in a way that maintains our core selves, balancing plasticity and stability to retain prior knowledge and adaptive habits while allowing new behaviors to be learned. Those core elements of selfhood are physically embodied in the structures of our brains, guiding, informing, and, yes, constraining our behavior, mostly through subconscious mechanisms. That does not mean we cannot consciously think about our motivations and beliefs and reasons for doing things. As we will explore in the next chapter, we certainly can and we certainly do, just not all the time. But it is simply a mistake to identify ourselves just with our conscious minds when, in any case, conscious mental activity also must be instantiated in some neural substrate.

FIGURE 10.4. The self. **(Left panels)** A person with the kind of free will envisaged by some—completely free of the influence of any prior causes—could do whatever, but not *whatever they want* because they wouldn't have any wants. They would have no reason to do anything in particular or anything at all, in fact. They would have no continuity as a person through time and would not really exist as a self at all. **(Right panels)** Our goals and knowledge and commitments provide us with enduring reasons for action and the ability to act on them. "We" only exist as selves extending through time.

Free will skeptic Sam Harris has argued that the compatibilist view put forward by philosophers like Daniel Dennett—where we are constrained to act according to our own reasons—amounts to saying that "a puppet is free as long as he loves his strings."[2] To me, this misconstrues the nature of the self. *Selfhood entails constraints*: they are what selves are made of. The puppet is made out of strings—if you removed them, there would be no puppet left.

2. Sam Harris, *Free will* (New York: Free Press, 2012), 10.

# 11

# Thinking about Thinking

It is time to finish our evolutionary story. We left off, in chapter six, with the emergence of animals equipped with sophisticated behavioral control systems that allow them to choose their actions, based on reasons. Those reasons derive from natural selection and, through learning, from the experiences of individual organisms. The philosophical discussions over the subsequent chapters made the case that such organisms can rightly be considered causal agents. That is, it is the organism *as a whole*, not only the physical happenings inside it, that is the cause of things happening.

But does this kind of agency get us all the way to what people understand as free will in humans? Or does it still leave unanswered the charge that we can do what we want but we can't want what we want? As free will skeptic Sam Harris asks (and answers), "Did I consciously choose coffee over tea? No. The choice was made for me by events in my brain that I, as the conscious witness of my thoughts and actions, could not inspect or influence."[1]

I argued in the last chapter that this essentially dualist framing is flawed in that it misidentifies *us* with just our conscious selves, in a given moment of time. But, beyond that point, the scenario outlined by Harris doesn't seem like an accurate picture of how we make decisions. It may describe our decisions regarding trivial things of no consequence, like choosing coffee over tea. But is this picture generally true? Are our

1. Sam Harris, *Free will* (New York: Free Press, 2012), 4.

motivations and intentions really always opaque to "us" (our conscious selves)? Do they indeed just bubble up to consciousness, leaving us free to choose, but for reasons we cannot consciously apprehend or comprehend and that we cannot intervene in?

## The Deluded Self?

This view of our conscious selves as merely the passive witnesses of our thoughts and actions has been seemingly reinforced by findings from neurology, popularized by writers such as Oliver Sacks, Michael Gazzaniga, Vilayanur Ramachandran, Antonio Damasio, and others. Many patients who suffered some kind of brain lesion—due to a tumor, stroke, head injury, or some sort of brain surgery—exhibit what is known as *neglect*. They seem unaware of some part of the world or unable to consciously report on something to which they previously had conscious access. In the case of split-brain patients in whom the major connections between their cerebral hemispheres are surgically severed to prevent the spread of severe epileptic seizures, one hemisphere may be unaware of things that are selectively presented to the other one.

What is striking in many such neurological patients is that they may still demonstrably respond to or act on information that is only available subconsciously. Moreover, when asked why they did something, they often *confabulate*; that is, they report reasons that are clearly made up. In split-brain subjects, for example, information can be presented to just the left or the right hemisphere, and the subject can then be asked to act on that information and then tell you why they chose a particular option. Because language faculties are lateralized to the left side of the brain, only that side can verbally report on its reasons. But in split-brain subjects, the left side doesn't know what the right side saw, so when the person does something based on that information, the left side of the brain is actually at a loss as to the reason.

However, very consistently, rather than just saying that they don't know why they did something, split-brain patients tend to make up some story that kind of fits what the left brain does know. For example, when one such subject, described by Michael Gazzaniga, was asked to pick objects with either hand that were relevant to pictures shown to each hemisphere, they picked a chicken to go with a chicken claw and a snow shovel to go with a snowy scene. But when the left hemisphere was asked why the person (really the right hemisphere) had picked the shovel, they said (not having experienced the snowy scene) that it was "to clean out the chicken shed."

These kinds of findings have reinforced the notion that we are not actually consciously directing our own motivations, moment to moment: we are not really reasoning through what to do but rather are engaging in some kind of post hoc interpretation or rationalization of the actions directed by our brains. In short, we're telling ourselves stories in a desperate attempt to contextualize our own actions. The implication is that we're doing this all the time, not only in the context of some brain pathology.

Indeed, a broad stream of psychological research literature claims to show that we are all prone to being primed by all kinds of external influences that shape our goals and behaviors without our knowledge and that, when challenged to explain why we behaved or chose in a certain way, we often make up stories, similar to patients with brain pathologies. For example, students working on language puzzles that included achievement-related words were—supposedly—found to perform better on a subsequent puzzle than other students who had been exposed to neutral words. Similarly, people could supposedly be primed to behave more competitively if, on entering an office, they saw a leather briefcase on the desk, or they could be primed to speak more quietly if they saw a picture of a library on the wall.

Much has been made of these kinds of findings; for instance, they have been taken as evidence supporting the Freudian view that we are driven by subconscious motives and goals of which we are unaware. Under this view, "we" are not really in charge at all: our brains decide and we are just left scrambling to explain ourselves. In his bestselling book *Thinking, Fast and Slow*, Daniel Kahneman leaned on these kinds of studies to argue for the pervasive influence of implicit priming on human behavior, writing, "Disbelief is not an option. The results are not made up, nor are they statistical flukes. You have no choice but to accept that the major conclusions of these studies are true."

Well, it turns out that the validity of this literature on psychological priming has indeed been called into question. The seminal publications are characterized by small samples, questionable statistical methods, and, in some cases, outright fraud. Recognition of these problems and of a systematic bias (shared by many fields) for only publishing "positive" findings led researchers in the field to attempt systematic replications of these experiments, most of which did not succeed in repeating the initial findings. These developments led Kahneman, previously the most vocal of proponents, to publicly disavow the studies on which his arguments were based. It turns out that many were made up and many of the rest do seem to have been statistical flukes.

The idea that we are not really in charge of our actions thus does not seem well justified. Even if we can sometimes be primed by external factors, this does not mean that we never make our own conscious decisions for our own reasons. And just because patients with brain lesions draw and report inaccurate, even fanciful, explanations of their behavior under pathological circumstances, this does not mean that that system is not well suited to draw accurate explanations under normal circumstances, when all the relevant information is available and all the circuits involved are working normally.

By analogy, just because our visual system can be fooled by carefully crafted optical illusions into making inaccurate inferences about what is out in the world does not mean that it doesn't normally make accurate inferences. If it did not, it would not be of much use, given that we rely on the accuracy of those inferences to get around in the world. Optical illusions are useful in highlighting some of the underlying processes that normally do this work. In the same way, findings from neurological patients highlight the existence of systems that interpret our actions. But they do not imply that these systems are always inaccurate or engaged only after the fact, as opposed to during the decision-making process while we're reasoning about our reasons. The same argument about usefulness also applies here: these systems would not exist if they consistently gave us false accounts of our own motivations.

It would seem strange if we were walking around constantly mystified by our own behavior, unable to explain our own actions, even to ourselves. That doesn't seem to fit with our common experience. In fact, we can and do often explain our actions not only to ourselves but also out loud to other people. Much of our everyday conversation is taken up with discussing our respective goals and desires and intentions as explanations for completed or intended actions. It's vitally important, for humans as an ultra-social species, that we understand each other in these ways—that we make sense of patterns of behavior by reference to underlying motivations. And it's just as important for regulating our own behavior that we understand our own motivations and can make accurate judgments about them.

The most recent stages of evolution along our lineage equipped human beings with a set of cognitive resources that grant us precisely this ability to reason about our own reasons. This gives humans an extra level of self-regulatory control—the faculty of metacognition—that we can and do use to think about and adjust our own motivations.

## Becoming Human

We talked in earlier chapters about the idea of *cognitive depth*; for example, if a nematode worm could be said to be thinking, it's certainly not thinking *about much*. It can only detect very immediate olfactory or tactile stimuli and has very few levels of intermediate processing between perception and action. It may integrate a few signals at a time and can do simple forms of learning, but it doesn't create much of a map of the world or its own self and doesn't do any kind of long-term cognition. It inhabits the here and now.

For bigger creatures, the move to land made vision and hearing much more valuable and led to the evolution of systems to map objects in the world at greater distances. This conferred value on the ability to remember and plan over longer time horizons, necessitating systems for adjudicating between short- and long-term goals. It gave organisms *much more to think about*. And it made developing better neural resources more and more valuable—at least in our lineage that adopted the *cognitive niche*. The smarter primates and early hominids got, the more advantageous that intelligence became. Bigger brains paid for themselves.

These advantages do not derive only from an increase in raw computing power; they come from the structured addition of new levels capable of abstracting information and *thinking about new things.* This addition may, in fact, happen naturally: as the cortical sheet expands, there is a tendency for existing areas to split into two, creating new areas that can act as new levels of the processing hierarchy. We saw in the visual system how more and more complex and abstract information is extracted and represented along the hierarchy of cortical areas—starting with simple lines, then shapes, then objects, and then types of objects, like faces or scenes. These object representations are linked to memory systems to enable recognition, activating a schema of known properties and associations and connecting percepts to concepts.

Now we are in a world of ideas. As the cortex expanded and new areas arose beyond the perceptual systems, it became possible for humans to think about more and more abstract categories of objects and types of causal relations between them. Those categories are built from combinations of simpler features, providing an efficient way to store and map knowledge in a nested hierarchy. Once you bind some lower-level features as defining properties of a category of thing, you can "name" it (neurally speaking, if not actually linguistically) and manipulate it as a higher-level concept. For example, "has feathers, walks on two legs, can fly" defines the category of "birds," and once you know about birds you can think all kinds of new thoughts about them, continuously adding to your web of knowledge at that higher level.

Different sets of the same lower-level features can be used as components of other concepts; "can fly" includes insects and bats, for example. The boundaries between categories can be sharpened by reinforcing one attractor state and inhibiting other ones representing similar concepts, which can be accomplished through the kind of neural dynamics we discussed in chapter nine. With these kinds of latent categorical representations configured into the system, you can imagine a concept—bring it to mind—by activating top-down the web of activity that represents its collective features. If I ask you to think of a bird, you can do that, and all the birdy properties come along for free.

With these kinds of concepts in place, we can build frameworks of causal relations. The world is made of distinct entities that we are naturally attuned to recognize (like individual birds), but these entities are often joined in systems of interacting components that have regular, observable dynamics (like mating pairs or families or flocks of birds, or whole populations in an ecosystem). At any of these levels, we can treat the individual or the family or the population as a *whole*—a unit of analysis—and try to figure out its dynamics at that level. In doing that, we may find that the family dynamics in some birds—involving extended parental care of the young, for example—mirror those in other species, including our own. Now we've learned something much more general that we can apply in novel situations.

By being able to think at this level, we turn isolated elements of knowledge into a more general *understanding* of how the world works, something that artificial intelligence still struggles to do. And we can deploy that understanding in directing our own behavior, even in ostensibly novel situations. We can combine these nested hierarchies of concepts and maps of causal relations and system dynamics in new, creative ways within this abstract cognitive space and thereby engage in open-ended, model-based reasoning. *We can imagine things*. In effect, we can mentally simulate a model of the world and "run" it in fast forward, predicting and evaluating outcomes of various actions over short and long timeframes.

Our ability to model the world in this way gives us unprecedented control over our environments. When faced with some problem, we have the ability to see the bigger picture by taking into account a wider context and a longer time horizon. This means we can avoid getting stuck in *local optima*—the quickest, easiest solution to a local problem—and instead optimize for global parameters. We can think strategically, not just tactically. However, the exercise of true control requires one other crucial element: we must also model ourselves. Not just where our bodies are in the world or a record of the actions we're taking so we can adjust perception, as other animals do, but model our own decision-making machinery.

## How Am I Doing?

A key factor in the regulation of your behavior is to assess how you're doing. We discussed in previous chapters the many ways this assessment is accomplished in other animals. These ways include monitoring actions as they are being executed and making any necessary adjustments to accomplish the goal. In addition, *instrumental* learning from the outcomes of actions involves having a map of current goals and intended actions and comparing the state of the world and the organism itself after an action to determine whether it was completed successfully and if it turned out well. In the simplest versions, reward signals can reinforce a specific behavior in a specific situation, and punishment signals will do the reverse. But there are more sophisticated aspects to this kind of learning, which require additional sorts of evaluation.

If an action didn't turn out well, for example, is that because I had bad or incomplete information? Was some source unreliable? Do I need to look for more information? Do I need to update my model of the world to correct apparently wrong beliefs? Did I just not execute the action properly? Or did I have good information but a bad strategy? Should I try something else next time? Maybe I was just unlucky on that occasion—was some random element at play that I should ignore? Animals that can distinguish between these possibilities and adjust their world model appropriately will have greater predictive power of their environments and more adaptive behavior in the future.

One type of signal that is crucial in making these kinds of evaluations is certainty, as well as its correlate, confidence. This signal provides meta-information: information *about your information.* If, for example, some set of neurons is representing a perceptual belief—that there is a lion in the bushes, let's say—it is often very useful if it also carries some information about how certain the belief is. Maybe there really is a lion out there, maybe there isn't: to inform action it would be really useful to know how much confidence to have in that belief.

Hearing a rustle of leaves may be a less strong signal (especially if it's windy) than seeing the flick of a tail. But was it really a tail or just a long piece of grass?

In chapter eight I described one way to measure certainty. If a population of neurons has a number of possible attractor states that it can stably be in, each of which means different things, then another population of neurons that monitors the state of the first set can draw an inference about the certainty attached to the signal. If population A is very strongly driven into a given state, then over some short period of time it will mostly be signaling that one thing. But if it's being less strongly driven—if the incoming signals are more ambiguous—then it might oscillate between several possible states. The degree of any such vacillation can be measured by population B, if it can sample the activity of population A over some time period. Maybe it's signaling "X" 70 percent of the time and "Y" 30 percent, or maybe it's 90–10 or 50–50. Importantly, the first population can't "know" its own level of certainty at any moment—it requires a second population to monitor it and infer that parameter. In turn, that certainty may influence how the activity of the second population evolves by effectively allowing it to weight the information coming through the A channel (relative to any other inputs it may get).

Some process along these lines is probably happening throughout the brain at all levels of signaling in the nervous system. The resulting signals of certainty are an essential element in helping the system know how best to update its model of the world: there is no point updating based on an uncertain signal. They are equally crucial in allowing the system to update *its model of itself*, including its own processes of decision making. In many animals, the decisions that an animal makes also have some level of confidence attached to them—the level of belief that this was the right thing to do, under the circumstances. Behavioral and neural correlates of confidence can be detected and measured in experiments in which animals have to make some simple choice: How confident are they that the left arm of the maze really is where the food is, or that the signal really was red and not green, or that the odds of a payoff have or haven't changed? It's even possible to manipulate specific neural circuits in animals to tune the threshold of confidence required to execute an action (as opposed to waiting for more information) and thus make animals behave more or less impulsively.

Humans, of course, have all that metacognitive machinery. But we also seem to have a second-order level that makes these metacognitive machinations explicit. This means they are *consciously accessible* and, thus, reportable to other people. That's a huge benefit in a species like ours, where other human agents are the most important elements in our environment. But, more immediately, it means that our beliefs and goals and intentions, along with these second-order signals of certainty, can be collectively inspected and considered in a common space, with a common currency, and reconfigured if necessary. We really can reason about our reasons.

As Douglas Hofstadter put it (or had his fictional character, the Tortoise, put it):

> When an OUTSIDER ascribes beliefs and purposes to some organism or mechanical system, he or she is "adopting the intentional stance" toward that entity. But when the organism is so complicated that it is forced to do that with respect to ITSELF, you could say that the organism is "adopting the AUTO-intentional stance." This would imply that the organism's own best way of understanding itself is by attributing to itself desires, beliefs, and so on.[2]

Being able to inspect our own reasons, tagged with meta-information on certainty, we can better judge how to update our models after some action. With a first-order model, animals can realize they made the wrong choice and should do something else next time. With our higher-order models, humans can realize they made that choice *for the wrong reasons* and should *think something else* next time. We can learn what we should pay attention to in a given circumstance—what are the most relevant factors and the most reliable sources of information. We can do better in the future by learning what to learn from.

2. D. R. Hofstadter, Who shoves whom around inside the careenium? Or what is the meaning of the word "I"? *Synthese* 53 (1982): 215.

## Prefrontal Cortex and Metacognitive Skills

Our increased powers of cognitive control came with the expansion of the cerebral cortex, particularly the regions at the front of the brain collectively known as the prefrontal cortex. Prefrontal areas are present even in rodents, but they expanded hugely along our lineage, making up about 30 percent of the cortex. And they also became more complex, with many more identifiable discrete subareas (see Figure 11.1). The prefrontal cortex is particularly well suited to play a high-level role in monitoring and guiding behavior because it has connections to and from all other areas of the cortex and to the extended circuitry of action selection in the thalamus, basal ganglia, and midbrain regions that are involved in evaluating goal progress and action outcomes. This gives the prefrontal cortex a bird's-eye view of the distributed and diverse processes of decision making.

Prefrontal regions seem to be required for the aspects of cognitive control that we recognize as *acting rationally*—the abilities that, since the time of Aristotle at least, have been understood to set humans apart from other animals. These include planning over long timeframes, coordinating among multiple competing goals, inhibiting actions that otherwise would be favored, monitoring progress, maintaining information relevant to current tasks in working memory, assessing certainty, directing learning to the most reliable sources of information, and switching among goals when conditions change or new information becomes available. Collectively, these processes involve using our intellect to override our instincts or habits or to get us beyond apparently optimal local solutions to enable more globally optimal action.

Psychologists have developed a number of fiendish tasks to test these abilities. One famous one—the Stroop task—involves reporting the color of ink that a word is written in. That doesn't sound too hard, does it? If you see the word "boat" written in blue ink, your job is to say "blue." But sometimes the word is "green" written in blue ink: then your job is to ignore the cue given by the word itself and still report the color of the ink. That task is not so easy: most people struggle with it, having

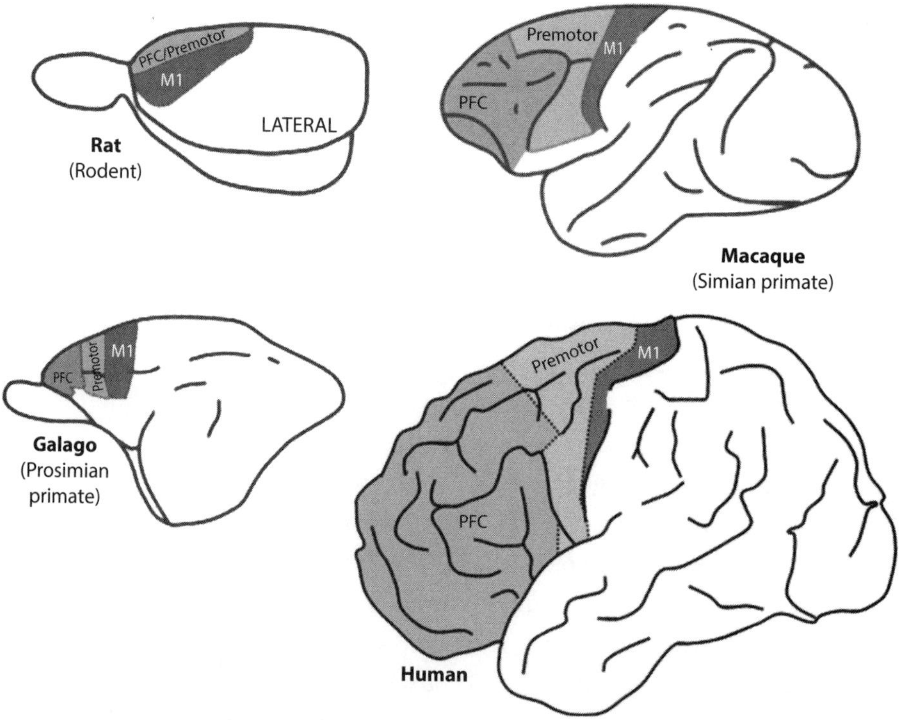

FIGURE 11.1. Prefrontal cortex evolution. The prefrontal cortex (PFC) has massively expanded along the primate lineage leading to humans, occupying a much larger proportion of the cortex overall, with a concomitant increase in the number of distinct subareas comprising it (not detailed). M1: primary motor cortex.

to slow down and actively think about it. That is, they have to inhibit the automatic response and apply conscious cognitive control to follow the rules of the task. The Wisconsin Card Sorting Test is another test of prefrontal functions; in this test, subjects have to sort cards depending on the shapes or number or colors of symbols on them, but the rules of the task are frequently varied so that subjects have to inhibit one learned strategy and switch to another to be successful.

People with damage to their prefrontal cortex struggle with these kinds of clinical tests, particularly in adhering to the rules, inhibiting incorrect responses, or switching strategies when appropriate. In their everyday lives, such patients often exhibit indecision, vacillation, or inappropriate perseverance. Depending on which parts of the frontal lobes are damaged, their symptoms may vary. Serious damage to the dorsolateral parts from a brain injury, for example, can lead to emotional flatness and behavioral inertia; such patients will not spontaneously initiate actions but, if prompted or guided to begin them, will continue doing them for far longer than necessary. Patients with damage to regions known as the orbitofrontal cortex may, by contrast, display very prominent emotional disinhibition and similarly disinhibited behavior. They may be easily distracted and show a childlike drive for immediate gratification, a tendency to act on every impulse, and an inability to foresee the consequences of their actions.

These symptoms highlight the crucial role of prefrontal regions in our ability to proactively regulate our behavior by balancing the capacity to keep focused on a task or long-term goal with the mental flexibility to adapt to changing circumstances. It is important to emphasize that although prefrontal regions sit at the top of a hierarchy in one sense, that does not mean they make all the decisions in isolation. Prefrontal regions do have a privileged position in receiving inputs from across the brain. But it is not accurate to think that all that information is just funneled to the prefrontal cortex where decisions are made, with orders then transmitted down to the rest of the brain. Instead, the process of guiding behavior involves reverberating conversations up and down the hierarchy, with different regions at different levels concerned with different aspects.

One of the functions of the prefrontal regions seems to be to keep certain information "in mind" while figuring out what to do or while executing some action or goal-driven behavior. The details of all that information are not necessarily *duplicated* within the prefrontal regions, however; prefrontal areas may instead send signals down to other regions of the brain that promote the active maintenance of the relevant representations locally, thus holding them in *working memory*. Prefrontal regions may not know everything themselves, but they know who to ask.

Armed with that information and able to operate on it in a common space, prefrontal regions then send descending signals to the parts of the brain involved in action selection. However, those signals do not simply impose activity on the other brain regions; instead, they seem to *bias the ongoing competition* among possible goals and actions that is playing out across these regions. Thus, the expanded prefrontal cortex in humans did not supplant or supersede all the decision-making machinery that exists in other animals: it just adds another level on top of it. Those biasing signals can inhibit some actions that might be taken habitually while promoting ones that are not immediately optimal in the service of longer-term goals.

In a classic paper proposing a theory of prefrontal functions in cognitive control, Earl Miller and Jonathan Cohen give the example of an American tourist in England, who has to exert conscious cognitive control to remember to look right when stepping out into the road, rather than left, inhibiting the habit of a lifetime. This is effortful and easily forgotten, especially if one is tired or distracted. But over time, if our American moves to England, then, as a result of learning the new regularities of the environment, the new behavior may become more habitual. When that occurs, the prefrontal regions may no longer be called on for oversight.

These ideas fit with theories of Daniel Kahneman and Amos Tversky about System 1 and System 2 thinking, mentioned in chapter ten. System 1 is fast, automatic, largely subconscious, and highly efficient—if you know what you're doing. System 2 is slow, effortful, largely conscious, and inefficient, but crucial when the situation is novel or ambiguous and there is no single "correct" answer. Decision making under those conditions may entail a multilevel optimization process, weighing different goals and strategies against each other, continuously monitoring the situation, and actively learning from the experience. With such learning, new behaviors become more routine and automatic, gradually switching to System 1. But when the need arises, System 2 can still be called on, in an instant, if we need to consciously take the wheel again. (Note that these supposedly separate systems really represent a continuum from more automatic to more deliberative action.)

## Consciousness

It may seem odd that I have not yet addressed consciousness much. This whole chapter is about conscious cognitive control, but we have not yet examined what role the "conscious" part plays in it. Consciousness is a multifaceted phenomenon, and it is important to take care to dissociate its various elements. The word is variously used to refer to the capacity of self-awareness generally (as opposed to *non*-conscious entities), actually being awake and aware of the world in a given moment (as opposed to being *un*conscious), or awareness of specific sensations or percepts (as opposed to them being *sub*conscious). For the question of cognitive control, it is the last usage that is most relevant (taking the other two as given).

How does it help with cognitive control for us to be conscious of some mental states but not others? You might think that it would be better to be aware of everything that's going on in our minds; surely, that would give us more information and more control. But it's easy to see that that would be overwhelming. We do not need or want complete information for optimal oversight: what we want is the right information, at the right level. The key to control is precisely the selectivity of conscious awareness. We are configured so that most of our cognitive processes operate subconsciously, with only certain types of information bubbling up to consciousness on a need-to-know basis.

Conscious perception seems to require the activity of prefrontal regions. It is possible to set up systems in the lab, for example, where visual stimuli are presented extremely rapidly, just on the threshold of conscious detection. When this is done, subjects sometimes report seeing the stimulus and sometimes not. By recording or imaging the brain, we can see that the areas of the cortex devoted to early visual processing are activated similarly in either case. What differs is that prefrontal areas also become active—slightly later—in cases where the percept is consciously perceived. In either case, the brain has detected the stimulus, and it can be shown that even subliminally presented stimuli affect responses to subsequent stimuli through perceptual priming effects. But it is only in the case of conscious perception that the subject reports

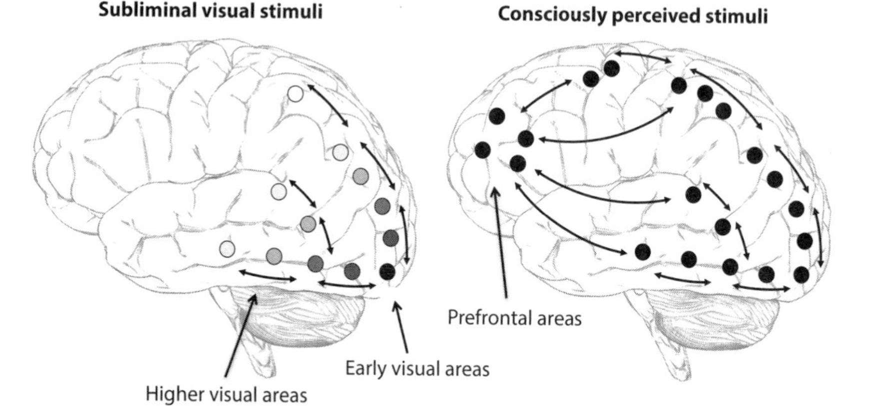


FIGURE 11.2. Consciousness. Visual stimuli can be presented in such a way that they only sometimes reach conscious awareness. EEG recordings show that early visual areas respond strongly, regardless of whether the stimulus remained subliminal (**left**) or was consciously perceived (**right**). Higher visual areas, by contrast, are only weakly activated by subliminal stimuli. Full conscious awareness is associated with slightly later "ignition" of a much wider set of frontal areas, with activity reverberating back down the hierarchy to visual areas.

what the stimulus was. The "ignition" of the prefrontal circuitry thus indicates that the percept captured the attention of the subject and entered consciousness (see Figure 11.2).

Crucially, this selective filtering of what makes it to the level of consciousness is itself an actively controllable process. First, our subconscious is continually monitoring things in the environment that might require our attention, from a flicker of movement in your peripheral vision to someone saying your name in a nearby conversation at a party. Beyond that, we also can choose what we want to admit to the arena of conscious thought—what we attend to and hold in working memory—or what we *think about* at any moment.

A classic experiment by Christopher Chabris and colleagues dramatically illustrates how strongly top-down attention restricts access to consciousness. In this experiment, subjects are asked to watch a short video of six people—three in white shirts and three in black—who are passing a basketball back and forth while meandering in and out of each other's paths. The task is to count how many passes the people in white shirts make to each other. This isn't particularly hard to do, but it does require paying close attention to the relevant stimuli while ignoring others. Exactly how much ignoring is being done is revealed when the subjects are later asked, "Did you see the gorilla?" In the middle of the video, a person in a gorilla suit calmly walks into the middle of the moving players, waves to the camera, and walks off. Most subjects report they were completely unaware of this interlude and express disbelief that it could have happened at all, until they are shown the video again. When they are not selectively attending to the task, the gorilla is, of course, perfectly, hilariously obvious.

According to one prominent theory, consciously exercising cognitive control thus involves *selectively* bringing together information about disparate aspects of the world, the state of our bodies, and our current cognitive processes with metacognitive signals of certainty into a common *global neuronal workspace*. This allows us to operate on all those parameters and to send signals back down to optimally organize behavior. This centralized, high-level oversight is the best way to effectively navigate our complex, dynamic, ambiguous environments; balance goals; and choose among competing options. At the same time, it also allows us to optimize our own cognition and learning.

This does leave unanswered some troubling questions, however: Why does it feel like something? How can such integrative high-order cognitive processing give rise to *conscious experience*? And does that experience itself have any functional role? These questions actually apply to the idea of subjective experience more generally. We recognize in sentient animals the capacity to experience sensations. This ability goes beyond just detecting and responding to stimuli, as a robot would, but exactly how is hard to define. There is certainly a sense that some animals are *more sentient* than others and that there may be very different qualities of subjective experience across the animal kingdom. What distinguishes humans is sapience (hence the name). We don't just have sensations: *we have thoughts*.

As to how such conscious mental experiences come about, well frankly, we don't know. I discussed in chapter five the idea that animals that move around in the world must, at the very least, model their own existence and their own actions to enable them to distinguish self-caused from externally caused perceptual changes. This necessarily entails a subjective perspective, which may in some way anchor subjective experience, however basic that may be. In a similar vein, perhaps our modeling our own cognitive processes just necessarily produces conscious, first-person, *mental* experience. Thinking about our thoughts entails identifying them as *our* thoughts. That this is an active process is suggested by conditions where it seems to break down, as in schizophrenia, where our thoughts can be ascribed to somebody else. Maybe anything with that kind of recursive architecture would be conscious to some degree.

On the question of what the *conscious* nature of those experiences is good for, one huge benefit—perhaps a truly transformative one—is that it enables us to *imagine*. We're not aware of the firings of all our neurons or the patterns of activity in every brain region. But we are aware, at any moment, of *the meanings* carried by some select subset of those patterns: we mentally experience sensations, percepts, beliefs, goals, intentions, emotions, feelings, moods, thoughts, and ideas. Being able to operate on them as cognitive objects gives humans a unique ability—we can manipulate ideas in truly novel, creative ways; turn them over in our minds; test them out in mental simulations; and explore new avenues of thought disconnected from the immediate consequences of action. What evolved initially as an extra level of our control architecture, allowing us to model and regulate our own cognitive operations, may thus have been just the thing that freed our minds.

The other striking benefit of conscious experience relates to social communication and cognition in two ways. First, being able to identify our own mental states also means we can tell someone else about them. We don't have to think alone. In ways that we do not fully understand, the evolution of human consciousness, of language, and of culture all probably relied on and reinforced each other. What started as simple sounds and gestures to convey hunger or anger or fear or danger evolved into our uniquely human capacity for language, allowing us to express more complex mental states. By sharing thoughts and ideas, we could build on each other's experience, accumulate hard-won knowledge, and drive a cultural explosion not seen in any other species.

In addition, the evolution of our ability to model our own thoughts probably went hand in hand with the ability to model *other people's thoughts*. It's easy to see the selective advantages that would arise in individuals who are better able to intuit the beliefs and desires and motivations of other people. Whether competing or cooperating, knowing what your fellows are thinking and feeling is hugely beneficial in predicting, anticipating, preempting, or manipulating their behavior. As we evolved into the ultra-social, ultra-cooperative, obligately cultural species we are, *other minds* became, by far, the most important things in our environments.

## Mental Causation

This chapter began by considering the claim that we are merely passive witnesses to our thoughts and lack insight into our motivations. To the contrary, the evidence presented here shows that humans possess a highly developed set of neural resources devoted precisely to metacognition, introspection, imagination, and conscious cognitive control of our behavior. We think about our thoughts, and we reason about our reasons. In fact, much of our waking mental lives is taken up with such introspection, and when we're not thinking about our own thoughts and reasons, we're often thinking about those of other people.

These capabilities really do give us conscious control over our own behavior. They do not always have to be engaged in the moment of action, however. As discussed in the last chapter, many of our actions are guided by habits and heuristics and other learned adaptations to our environments. In addition, much of the heavy lifting of decision making is often consciously done ahead of time. We make commitments to extended courses of action, we adopt policies that apply across many future situations, and we plan for contingencies so that, when they arise, we simply have to execute the action we previously consciously decided to do.

But we also retain conscious oversight and momentary control that can be exercised when the need arises. These abilities rely on prefrontal regions that bias the competition among possible actions, prioritizing some and inhibiting others. Of course we don't know all the details of how this is accomplished, but the general picture outlined here is well supported by experimental evidence; it seems to provide a perfectly naturalized account of human agency or even what most people understand as free will. Why then are so many people—including many neuroscientists—so inclined to argue that we *do not* actually make choices and that we're not in charge of our own actions?

This skepticism seems partly due to the enduring intellectual legacy of French philosopher and mathematician Rene Descartes, which has shaped the Western scientific tradition. One of Descartes's most famous ideas is that the world is made of two very different types of substance: the physical and the mental. This *dualist* position gets around having to explain how physical stuff can produce immaterial things like thoughts by simply positing that thoughts occupy a kind of parallel realm of the mental. The problem with this idea—pointed out by some of Descartes's contemporaries, such as the astute and wonderfully titled Elizabeth, Princess of Bohemia—is that it does not explain how the physical and the mental realms can interact. They clearly seem to, because thinking about doing something can indeed lead to us doing it—physically moving our bodies and things in the world—but how? Descartes did not have a good answer to this question (though he did propose a route of communication through the pineal gland, for no particularly good reason).

You would think we would have moved on by now, after four hundred years, but it seems we still get hung up on a version of the same question: *How could having a conscious thought move physical stuff around*? Doesn't that somehow violate the laws of physics? It seems to require a mysterious form of top-down causation in which the mental pushes the physical around. But this apparent mystery only arises if we think of the mental as some realm of free-floating thoughts and ideas. It's not a question of whether immaterial thoughts can push around physical stuff. *Thoughts are not immaterial*: they are physically instantiated in patterns of neural activity in various parts of the brain, which can naturally have effects on how activity evolves in other regions. There's no need to posit a "ghost in the machine"—you're not haunting your own brain. The "ghost" is the machine at work.

However, if this gets us away from dualism, it seems to push us into the arms of a materialist view that threatens to deny the mental content—the meaning of those thoughts—as having any real causal power. If they are *just* patterns of neural activity, which through physical means have some effects on the unfolding of neural activity in other parts of the brain, then aren't we back to a reductive view of ourselves as just complex machines?

The way out of this conundrum is to realize that thoughts are not *just* patterns of neural activity: *they are patterns that mean something.* As we explored in chapter nine, it is the meaning of the neural patterns that drives the unfolding of neural activity in the brain—the low-level physical details are often incidental. So, although thoughts must be instantiated in some pattern of neural activity, such patterns only have causal power in the system by virtue of what they mean. Some subset of those meanings are consciously apprehended. When they are, that *just is* you thinking. And when you are manipulating those cognitive representations and consciously working out what to do, that *just is* you deciding. It's not an illusion or an epiphenomenon or a post-hoc rationalization. Mental causation is a perfectly real and naturalizable phenomenon (see Figure 11.3). There is nothing spooky at work, but neither can it be reduced to mechanism.

To sum up, the evidence I presented in this chapter argues strongly that we really are aware of our own motivations and do have the capacity for conscious cognitive control over our actions. There is, however, one final wrinkle in this story, which is that some of us may have this capacity to a greater extent than others.

## Individual Differences

The most obvious variation in the faculties of self-control is seen across the lifespan. Babies are clearly not born able to use reason to control their behavior: they are entirely driven by their basic physiologi-

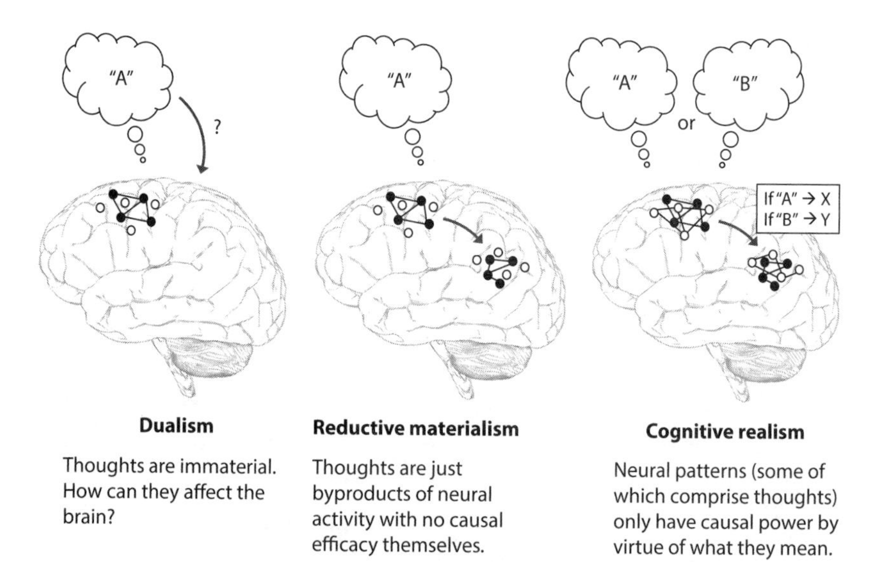


FIGURE 11.3. Mental causation. **(Left)** A dualist model suggests that "immaterial" thoughts have some causal power but cannot explain how they can affect the physical substrate of the brain. **(Middle)** Reductive materialism admits that thoughts are associated with certain neural patterns but does not give causal power to the meaning of those thoughts. **(Right)** A more naturalized position (which I loosely call "cognitive realism") sees neural patterns, some of which comprise thoughts, as having causal power in the physical system solely by virtue of what they mean (i.e., based on the criteria configured into the system itself).

cal urges (or, more accurately, their parents are). It doesn't take that long for infants and children to start engaging in something you could call reasoning, but the ability to behave rationally—to control their urges, inhibit their impulses, and regulate their emotions—takes years to develop. It typically involves a lot of explicit instruction and reinforcement by parents, teachers, and other caregivers, as well as learning from peers.

At the same time, there are also clearly individual differences in how these regulative capacities emerge, illustrated, for example, by the famous *marshmallow test.* In this test, devised by Walter Mischel and colleagues in the 1970s, young children (from three to five years old) were brought into a room and sat at a table with some treat, like a marshmallow, placed on a plate in front of them. The researcher then says he or she has to leave the room for fifteen minutes and the child can eat the treat if they want to, but if they wait until the researcher returns, they will get two treats. The idea is to see how much these children could delay gratification in the service of a greater reward. You can see in videos of these kinds of experiments the torture as the kids tried desperately to distract themselves from the yummy marshmallow, covering their eyes, looking at the ceiling, singing to themselves, or playing games with their hands and feet to keep from succumbing to temptation. Not surprisingly, some kids were able to wait longer than others.

What was more striking was the result of follow-up studies of those same individuals two to three decades later. These studies examined life outcomes, financial parameters, level of education, body mass index, smoking behavior, and other factors, as well as assessing the subjects' capacity for self-regulation through questionnaires. The headline finding—which became a "fact" in the public perception—was that the ability to delay gratification on the marshmallow test as a young child was predictive of these later life outcomes, with children who waited longer being more successful, healthy, and well adjusted. More recent studies, some by the same authors, called these conclusions into question, however, indicating that small samples, variation in test procedures, and the effects of confounding factors such as socioeconomic status may have led to the apparently positive correlations.

Nevertheless, even if behavior on this one fifteen-minute test at age three to five does not predict one's future life trajectory, it does illustrate a kind of variation in the degree of self-control that we also readily perceive in adults. In our everyday lives, we recognize that some people are characteristically impatient, impulsive, short-tempered, emotionally labile, feckless, reckless, and easily distracted, whereas others are controlled, temperate, patient, prudent, perseverant, emotionally stable, focused, and disciplined. It's not a surprise that our modern cultures reward the more stable, controlled characters who are better able to resist short-term impulses and carry through on longer-term goals. It is interesting too that these character trait terms are not neutral but have an inherent value judgment attached to them, with the ones reflecting higher self-control clearly perceived as morally "better," presumably because self-control is an essential element of prosocial behavior.

The way these characteristics develop most likely reflects an interaction between innate biological differences and experience. If the capacities underlying self-regulation and conscious cognitive control are in some sense written in our DNA as humans, then we should expect to see them vary across the population according to genetic differences. And if they are like other psychological traits, they should also be affected by inherent variability in the processes of brain development. To investigate these issues, it is necessary to define and measure some underlying traits. This is part of a wider project to try and understand the dimensions of self-control along which individuals can vary.

Analyzing self-control has been attempted in a number of different ways, including looking at performance on various tasks carried out in laboratory settings, analyzing profiles of traits assessed by questionnaires, and measuring real-world outcomes that seem to reflect self-control. Ideally, we would be able to relate these different levels—discrete cognitive tasks, real-world behavior, and life outcomes—but the reality remains much messier than that.

People do show relatively stable individual differences in performance on tasks like the Wisconsin Card Sorting Test, mentioned earlier, and a host of others designed to tap ostensibly separate elements of cognitive control. One prominent theory defines three such elements: inhibition of habitual or locally optimal actions, updating of information in working memory, and shifting of goals to reflect new information. These factors are not really so cleanly dissociable, however, nor are they exhaustive; indeed, there is a partial overlap between the *executive functions* defined on these tasks and intelligence more generally. The important point, however, is that variation in performance on such tasks is indeed partly heritable and partly influenced by nongenetic variation that is not related to the family environment.

As it happens, performance on these tasks does not correlate well with either reports of real-world behavior or with life outcomes. A recent study by neuropsychologist Russell Poldrack and colleagues identified multiple factors in behavioral traits, including impulsivity, goal directedness, emotional control, risk perception, and reward sensitivity. They found that these factors were essentially unrelated to those underlying performance on tests of executive function. Although the traits measured on questionnaires did weakly correlate with a number of life outcomes, such as income, smoking, obesity, problem drinking, and mental health, the factors derived from the tasks did not.

The upshot is that performance on laboratory tasks measuring real-time executive function is partly heritable, but the relationship to behavioral traits of self-regulation—themselves also partly heritable—is not so simple or direct. This finding in no way undermines the idea that such traits have a biological basis. It merely suggests, not surprisingly, that the functions underlying conscious cognitive control and self-regulation are diverse and distributed and cannot be fully captured by abstract tasks in the lab. This complexity is reinforced by pathological conditions that impair self-regulation in diverse ways.

We discussed already how brain injuries to prefrontal regions can impair conscious cognitive control and behavioral self-regulation in different ways. There are also any number of psychiatric conditions that seriously affect these faculties, including attention-deficit hyperactivity disorder, autism, schizophrenia, bipolar disorder, obsessive-compulsive disorders, Tourette syndrome, dementia, alcoholism, and drug addiction. What is striking in surveying these conditions is the variety of ways in which self-regulative capacities can be impaired; for example, the inability to focus on long-term tasks, inhibit inappropriate actions, regulate emotionality, maintain relevant ideas in working memory, or resist the temptation of short-term rewarding stimuli such as drugs or alcohol. Some of these impairments reflect acutely pathological brain states, such as psychosis or mania; others are more trait-like but veer into deficits that affect real-world functioning.

These conditions are obviously very serious, and working out their underlying etiology and pathology is of crucial clinical importance. What is most relevant for our discussion in this chapter, however, is that these diverse types of individual differences provide another source of evidence that what we think of as free will—the ability to consciously monitor and control our cognitive processes, to reason about our reasons, and thus regulate our own behavior—is not some abstract metaphysical postulate but an evolved function, or suite of functions, with a very real biological basis.

# 12

# Free Will

We started in chapter one with some challenges to the notion that we are really in charge of our own actions. In our everyday experience we seem to be able to make choices, to decide things of our own accord, and select our actions—to be the authors of our own stories. But how could this really be true if our actions are determined by the way our brains are wired, or the firings of neural circuits, or the playing out of the fundamental laws of physics? These challenges pose a much more basic question: What kind of a thing are you? I argued over the course of the preceding chapters that you are a causal agent, a self with autonomy and continuity, capable of accumulating causal potential and exercising it in the world. In this chapter, we examine the implications of these claims and address this question: *Is this free will?* But first I review the main ideas of my thesis, the story of how agents evolved, and some of the themes and perspectives that help conceptualize and naturalize it.

## The Rise of Agents

The story of agency is really the story of life itself. For billions of years, no entities capable of action existed. But the emergence of life brought with it *activity*. Living things strive, actively, to keep themselves organized. They take in free energy and perform work to stay out of thermodynamic equilibrium with their surroundings. The barrier of a cell membrane gives even single-celled organisms a considerable degree of causal insulation and autonomy from the environment. They are not pushed around by every passing physical or chemical disturbance but are protected in their private bubble, selectively taking in materials and energy to keep themselves going. Organisms thus cannot be understood as static machines or instantaneous arrangements of matter: instead, they are patterns of interlocking dynamical processes that actively persist through time.

Over evolution, new tricks emerged to better enable organisms to persist, especially in the face of changing environmental conditions. Most obvious is the ability to move somewhere else. But where? Being able to move made sensation valuable; it meant that organisms could profit from *information* about what is out in the world. Specialized sensors evolved that could detect objects or substances or disturbances of relevance to the organism, and these sensations could be coupled to the motor system to guide behavior. Natural selection drove pragmatic couplings of approach or avoidance to or from different stimuli. The first glimmers of meaning and value inhered in these responses: approach or avoidance was *good or bad* for the organism, relative to its goal of persistence.

Organisms now had reasons for doing things. In a sense, those reasons are just wired into their biochemistry, but even for the simplest types of behaviors, it is still the whole organism that is acting. In single-celled creatures like bacteria or amoebas, receptor proteins for particular stimuli initiate signals that are transmitted to the motor machinery. But the organism is not a passive machine, just sitting there quietly waiting to be stimulated. It is constantly active both internally and in the world—it merely adjusts that activity to accommodate new information. Moreover, although it is possible in experimental situations to drive one pathway while holding other conditions constant—giving the impression of a dedicated, linear mechanism connecting stimulus directly to response—in reality these organisms integrate multiple signals at once, along with information about their current state and its recent history, to produce a genuinely holistic response that cannot be deconstructed into isolated parts.

Single-celled prokaryotes—bacteria and archaea—are amazing creatures, capable of transforming the planet and thriving in even the harshest environments. But they faced an energy barrier that prevented them from getting bigger and more complex. That barrier was lifted by the symbiotic event—the internalization of a bacterium by some archaeal species—that gave rise to mitochondria and, in the process, to eukaryotic organisms. These creatures could get much bigger and, crucially, could maintain a larger set of genes, enabling greater diversification of cell states. With the advent of multicellularity, these cell states became cell types. New types of selves emerged—ones with a division of labor between their components, with individual cells now working in the service of the whole organism and ceding reproductive duties to the germline. What was good for the whole organism was good for its parts.

More complex creatures emerged, colonizing and creating new niches, with expanded repertoires of possible actions. A system was then required to coordinate the movement of all the organism's constituent parts and select among actions. Muscles evolved, along with neurons to coordinate them, initially distributed in simple nerve nets. As evolution proceeded, the nervous system became more complex, linking sensory structures to muscles via intervening layers of interneurons. The meaning of signals became disconnected from immediate action, giving rise to internal representations—patterns of neural activity that do not just have pragmatic consequences but rather have semantic content: they *mean something* to the organism. The meaning of those internal representations is still grounded by inputs from the world and outputs through motor action—they just are a couple or more steps removed in both directions. Organisms with this kind of neural organization thus became truly cognitive.

When life moved to the land, new horizons opened up, literally. Vision and hearing became much more valuable as organisms could detect things that were far away. It started to pay to think over longer timeframes. More complex brains became a worthwhile investment. Organisms developed more and more levels of internal neural machinery, enabling them to extract higher-order information and to construct maps of objects in the world. They began to have much more *to think about*.

In parallel, neural systems for learning about things in the world also expanded, enabling organisms to build up knowledge about objects, properties, categories, contingencies, associations, statistical regularities, and causal relations. All that effort was worth it because that knowledge could be used to inform action. Mammals in particular began to develop much greater behavioral flexibility, with a richer, more diverse, almost open-ended repertoire of actions to choose from. That, of course, required a similarly sophisticated control system.

New control structures emerged—particularly in the expanded neocortex and interconnected forebrain regions—that provided extra levels of behavioral coordination and an ever-growing ability for individual organisms to learn from experience and adapt behaviors throughout their own lifetimes. These distributed systems integrate information on the individual's current state, motivations, short- and long-term goals, perceptual inferences, and beliefs of what is in the world, all interpreted in the context of stored knowledge and affordances and all in order to suggest some possible actions. These elements can then be weighed against each other based on predicted costs and benefits in a recursive process of biased competition. These complex systems enable organisms to simulate possible futures, enabling them to navigate a changeable and complex environment based on limited and ambiguous information, where there is often no obviously right choice or consistently optimal strategy. We are far beyond the concept of the organism reacting only to isolated stimuli: this is the organism deciding what to do for its own reasons.

The final stage in our evolutionary story—the admittedly egocentric story of our own lineage, that is—is the emergence of the ability to reason about our reasons. The expansion of the prefrontal cortex in primates, especially in humans, added more levels to our cognitive hierarchy. We moved beyond having a model of our bodies and a model of the world. Now we could have a model of our own minds—of our goals and desires and beliefs and of the certainty we should attach to them. We could operate on those ideas in a common cognitive space. Being able to reason about our reasons meant we could intervene in them: we could exercise top-down cognitive control, in the moment. And we could consciously make decisions prospectively for future actions, adopting policies and commitments to guide future behavior, in the process shaping our own character and actively constructing ourselves as we interacted with the world.

The conscious aspects of those processes remain mysterious. In popular discussions of the future trajectory of artificial intelligence, there is often mention of a *singularity* occurring: a moment when computers will achieve consciousness (usually followed by them deciding to do away with all the humans). This scenario remains far-fetched for computers, but clearly something like that must have happened in our own lineage. At some point in evolution, our internal models became so abstract and recursive that they gave rise to mental experience.

Perhaps the story I told here accounts for that evolution. Perhaps expanding the cortex in the way I described, adding new levels to the recursive hierarchy, necessarily leads to the system modeling its cognitive processes. And perhaps modeling one's own thoughts necessarily entails conscious mental experience. But it's also possible we're missing some other essential element that explains the apparent discontinuous jump from our closest primate relatives to humans. Of course, we should be careful not to forget that we are only seeing part of the picture. All the other species of hominids that existed along our lineage died out, so the discontinuity may appear more stark than it actually was.

Nevertheless, there still seems to be something left to explain to account for how we went from hairless, upright apes to the dominant lifeform on the planet. Several theorists, including Daniel Dennett, Kevin Laland, and Cecilia Heyes, for example, argue that biology was not enough, that this transition took a cultural revolution. Indeed, biologically modern humans existed for tens of thousands of years with apparently little change in lifestyle. Clearly we had the neural equipment to enable our eventual cognitive explosion, but some other forces were required to light the fuse.

The leading hypothesis is that the emergence of some kind of primitive language allowed humans to develop culture and to share hard-won knowledge with our fellows and, crucially, with our offspring, thus accumulating the abilities to control our environments over generations. This likely provided some positive feedback for our move into the cognitive niche, with each new advance opening up new possibilities in a self-amplifying process. Thus, through both biological innovations and cumulative cultural evolution, humans developed capacities for creative, open-ended, recursive thought and boundless imagination that truly set our minds free to combine and manipulate ideas in more and more abstract ways.

## Is This Free Will?

In the opening chapter, I discussed how a large part of the philosophical debate about free will is explicitly driven by concerns over moral responsibility. Threats to the idea that we can freely choose our actions also threaten to undermine the very notion that we may be held responsible for our actions and, consequently, the foundations of many of our social and legal systems. Much of the discussion in the philosophical literature is therefore aimed at justifying some understanding or definition of free will that would be sufficient to protect moral responsibility, presuming it needs defending from the perceived threat of determinism.

I purposefully did not start with a preconceived notion of what properties our will must have to qualify as "free," for this purpose or any other. Instead, I aimed to naturalize the underpinning concept of *agency*, with its core elements of purpose, meaning, and value, so as to arrive at an understanding of the properties, scope, and limitations of human decision making.

With that foundation in place, we can return to the timeless question, Do we have free will? We must first ask whether that question, as posed, is even coherent. The answer depends on your understanding of *free* and, less obviously, *we*. In philosophy, freedom in this context is typically defined as the ability to act *absolutely free from any prior causes whatsoever*. What would such freedom look like? Obviously, it means free from external coercion, which is fine, but does it mean you must be oblivious to the external situation? If you're taking the existence of something out in the world as part of your reason for taking some action, doesn't that constitute a cause? Any information that leads you to prioritize one action over others is, in fact, a constraint—a useful one but still one that reduces your options for action.

And if you're interpreting such information in the context of your memories and knowledge and understanding of the world and your motivations and goals, then those form another set of constraints—internal ones—that guide your behavior toward some purpose. To be free of such constraints would be to act randomly, pointlessly, on a whim, for no reason. You would, in fact, cease to be yourself; indeed, you would cease to be a self at all. Selfhood is defined by continuity through time—by maintaining a certain dynamic pattern of processes in the face of the thermodynamic pressure to take on any of the other, almost infinite sets of disordered arrangements those processes could adopt. Selfhood thus entails constraint. It *is only* constraint. The freedom to be you involves constraining the elements that make you up *from becoming not you.*

In humans, this activity includes maintaining the continuity of our psychological, biographical selves: all the memories and experiences and relationships; all the learnings, the habits and heuristics and policies; the commitments and projects and long-term goals; and all the dispositions that go with them. It means using the hard-won knowledge accumulated by your past self to guide action in the present moment in the service of your future self. If you're not doing that, you're not doing you. For these reasons, I take the absolutist definition of freedom to be incoherent in the context of the question whether we have free will. If you take "free" to mean absolutely free from any prior causes whatsoever, then you can have "free" or you can have "we," but not both.

The question of whether we have free will does not have a yes-or-no, all-or-none answer. Instead, we have degrees of freedom—an idea that is reasonably well captured, in my view, by a more commonsense understanding of the (still useful) notion of free will. That understanding entails, first, the ability to make choices—that we really can choose what to do. Our actions are not simply determined by outside forces because we're causally set apart from the rest of the universe to at least some degree. And, just as importantly, we are not driven by our own parts. Rather, we, holistically—*our selves*—are in charge.

The preceding chapters make the case that those criteria are met. Life itself entails a causal insulation from the rest of the world. Living organisms accumulate causal power over evolution and over their individual lifetimes. And we, like other animals, have a set of neural resources designed specifically to allow us to select our actions in the service of our goals; that is, to act for our own reasons. Those reasons inhere at the level of the whole organism, not its parts. The system acts as a unified whole (at least under nonpathological circumstances). Although it relies on subsystems and the workings of its physical components, its function cannot be deconstructed or reduced to those workings. The agent as a whole is deciding what to do. In humans, who have an added layer of conscious cognitive control, this seems to meet any reasonable, realistic criteria for free will.

Of course, there are many challenges to that view, which we also considered. At the most fundamental level is the idea of physical predeterminism—that there is and always has been and always will be only one possible future. This idea implies that the evolution of the entire universe and of everything that happens in it, including me writing this sentence and you reading it, was set from the Big Bang. This notion has always seemed to me fantastical and frankly absurd on its face—the kind of thing that only academics could take seriously. And it is undermined by most interpretations of quantum physics, as well as by information-based arguments that apply equally at classical levels, as discussed in chapter seven. The future is not defined with infinite precision: in contrast, things *become defined* through interactions in the time that we experience as the present. This is not an instant with an exhaustibly definable state but a duration of time during which processes evolve, until they are definitely resolved and move into the past.

The counterargument—that indeterminacy or randomness doesn't get you free will—also misses the mark. The idea is not that some events are predetermined and others are random, with neither providing agential control. It's that a pervasive degree of indefiniteness loosens the bonds of fate and creates some room for agents to decide which way things go. The low-level details of physical systems plus the equations governing the evolution of quantum fields do not completely determine the evolution of the whole system. They are not causally comprehensive: other factors—such as constraints imposed by the higher-order organization of the system—can play a causal role in settling how things go.

In living organisms, the higher-order organization reflects the cumulative effects of natural selection, imparting true functionality relative to the purpose of persisting (and all the more proximal subgoals that arise). These functionalities don't just emerge for free: they reflect design principles that are discovered by evolution because those arrangements work. The essential purposiveness of living things leads to a situation where meaning drives the mechanisms. Acting for a reason is what living systems are physically set up to do.

The other definition of determinism—that "every effect has a cause"—is, to my mind, tautological; that is, it assumes the thing it is trying to assert. If you define every event as an effect, then by definition it must have a cause. But is every event an effect? Some events (like radioactive decay, for example) seem to be truly random and thus should not be thought of as having a prior cause, at least nothing more specific than the universe being the way it is. The implicit idea—the reason that this formulation is put forward as a challenge to free will—is that all the causes must be located at the lowest levels of reality. But if the lowest levels are not fully predetermined, then this kind of reductionism need not hold: higher-order organization can be part of the cause of the things that happen. Agents themselves can be causes.

We also considered challenges from other quarters, particularly genetics, psychology, and neurology. The idea that we are not fully free because some prior causes that we did not control influenced the way our brains are configured has some merit to it. We are indeed somewhat constrained by evolution and our genetics and the way our brain happened to develop. All those elements did contribute to our psychological predispositions. But those predispositions do not *determine* our behavior, on a moment-to-moment basis, like the tuning of various circuits in a robot. Rather, they shape the ways in which we adapt to our worlds, choose and craft our environments, pursue our interests, and actively construct our character.

Nor are those motivations opaque to us. The evidence that neurological patients or subjects in psychological experiments do not always know why they did something, and sometimes engage in post hoc rationalization, does not mean we never have insight into our own reasons for action. We can and do reason about our reasons prior to and in the process of making a decision. We can and do exercise the capacity for conscious, rational control. Thus, although we cannot necessarily change our basal psychological predispositions, we are not slaves to them.

If free will is the capacity for conscious, rational control of our actions, then I am happy in saying we have it. Indeed, the existence of this capacity is brought home by the realization that some people have a greater capacity for this control than others. Babies, for example, are not born with this capacity, which entails a set of cognitive skills that must be learned and practiced. There is also inevitable variation across individuals in the diverse domains that underpin these cognitive faculties, contributing (along with social, cultural, and experiential factors) to the fact that some people achieve higher levels of rational control than others. In addition, these capacities can be reduced or even severely impaired in some individuals such as those suffering from compulsions, addiction, psychosis, depression, dementia, or many other forms of mental illness.

Moreover, just because we have this capacity does not mean we can always exercise it to the same extent. It can also be impaired by alcohol or drugs; by strong emotions like desire, rage, pain, jealousy, or grief; or even just by being tired or harried or distracted. Our free will is thus not some nebulous, spooky, mystical property granted to us by the gods. It is an evolved biological function that depends on the proper functioning of a distributed set of neural resources.

## Moral and Legal Responsibility

What are the implications of this position for our views on moral and legal responsibility? Can we still hold people responsible for their actions? Are they still deserving of praise or blame or reward or punishment? I believe that our views on these issues do not need to change. Despite the headlines proclaiming the death of free will, it remains stubbornly alive and well. Nothing in philosophy or physics or neuroscience or genetics or psychology or neurology or any other science undermines the idea that we do have the capacity for conscious, rational control of our actions.

Indeed, the fact that praise and blame are effective as learning signals—that they can lead people to alter their future behavior—supports the notion that we can consciously control not only our isolated actions but also our forward-looking behavioral policies. The social emotions that come with praise or blame—pride or guilt or shame—themselves act as powerful anticipatory signals that guide future behavior. We can even learn to avoid situations that we know impair our conscious cognitive control, like getting so drunk that we act inappropriately or recklessly, for example.

So there's no reason, in my view, not to continue to hold people responsible for their behavior in the ways we always have. This includes, as Daniel Dennett argues, ascribing to people the meta-responsibility to develop and maintain the capacity for self-regulation and conscious, rational control. The broad idea of moral responsibility is thus unaffected by discoveries from science that are revealing the neural and cognitive underpinnings of rational control.

As we gain ever-greater understanding of the ways in which the capacity for rational control varies and the conditions under which it can be impaired, there may be, however, some implications for how we assess legal responsibility. First, it is noteworthy that no trial lawyers ever call philosophers and theoretical physicists as expert witnesses to argue that physical predeterminism is true and therefore no one "could ever have done otherwise" in any situation. There is, however, a well-founded tradition of assessing mental or moral competence *across individuals*, taking age, insanity, intellectual capacity, or pathologies like brain tumors into account, along with circumstances that might temporarily impair cognitive control such as intoxication, delirium, or even sleepwalking. There is also a growing trend of people using more speculative genetic or neural evidence to argue that "their brains made them do it."

A 2016 study by Nita Farahany found that "neurobiological evidence is introduced in at least 5–6% of murder trials in the USA, and 1–4% of other felony offenses."[1] This kind of evidence is most commonly used

1. Nita A. Farahany, Neuroscience and behavioral genetics in US criminal law: An empirical analysis. *Journal of Law and the Biosciences* 2, no. 3 (2016): 485–509.

to assess competence or to argue for mitigation in sentencing, especially in death penalty cases. Indeed, a defense lawyer's failure to introduce neurobiological evidence is an increasingly successful ground for appeal on the basis of ineffective counsel. Most of these cases involve medical testimony, but some have drawn on brain scans or genetic findings from the scientific literature that supposedly indicate biological correlates of behavioral traits.

These have included, for example, genetic results showing that a defendant carries a genetic variant in a gene called MAOA that is supposedly associated with increased aggressiveness and antisocial behavior. There are some incredibly rare cases of people who have a mutation that completely knocks out the function of this gene who do show increased rates of physical aggression and violent crime. However, the claims of a much more general association with a *common* variant in the MAOA gene were shown to be spurious, arising from statistically flawed approaches. In general, there is an important difference between rare genetic mutations that cause clear pathology by themselves and the much more common genetic variants that contribute statistically, along with thousands of others, to variation in psychological traits across the typical range. The former may have important relevance to individual cases; the relevance of the latter is far more questionable. Indeed, the common genetic element with the largest known effect on propensity for violence is the Y chromosome, but we do not take its presence as exculpatory.

The same pattern holds for findings from brain scans. Some show obvious pathology, such as a traumatic brain injury or tumor, either of which can quite drastically alter behavior. Such findings are obviously relevant in court. But other types of imaging data submitted in court are based on much looser statistical patterns, not definitive single findings. For example, brain scan findings supposedly consistent with psychopathy or schizophrenia are sometimes used to augment the psychiatric evaluations of the defendants.

These claims rest on a scientifically much shakier body of literature, with many (often small) studies claiming that the size of some region of the brain or the thickness of some brain tracts differs between groups of people with some condition or trait versus those without. Typically, such claims have failed to replicate. In fact, there are no confirmed diagnostic brain imaging signs for any psychiatric condition nor any consistent, specific markers for traits like high aggressiveness or criminality. Even in situations where there are statistical trends produced by the study of large enough groups, such correlations do not imply causation or even specificity, nor do these findings amount to anything that could allow definitive statements about particular individuals.

There are echoes here of phrenology and physiognomy: the discredited ideas that the size of different brain regions (as supposedly revealed by the size and shape of bumps on the skull) or particular patterns of facial morphology, respectively, are associated with particular character traits, notably including "criminality." Nevertheless, modern advocates of *neurocriminology* argue that neuroscientific findings will only become more relevant in legal proceedings as we learn more about the brain bases of human behavior. For example, neuroscientists Andrea Glenn and Adrian Raine argue that neurobiological characteristics are not just relevant to judging culpability but also "could ultimately help to determine which offenders are best suited to specific rehabilitation programs and are more likely to re-integrate into society."[2] They even contend that genetic, neurobiological, or other markers could be used to predict future violent or antisocial tendencies in children—tendencies that might be preempted by treatment with medication or brain stimulation.

It is also likely that genetic data will begin to be presented more in court. Traits like aggressiveness, impulsivity, empathy, psychopathy, risk-taking, and many others that could feed into violent or antisocial behavior are all (like all our psychological traits) partly heritable. As we find more genetic variants associated with them, it seems inevitable that we will see presented in court *polygenic scores* for these traits—the profile in an individual of all the genetic variants associated with a given trait—at least as mitigating factors explaining why defendants were not

2. Andrea Glenn and Adrian Raine, Neurocriminology: Implications for the punishment, prediction and prevention of criminal behaviour. *Nature Reviews Neuroscience* 15, no. 1 (2014): 57.

really able to act differently than they did. As it happens, both neural and genetic arguments that a defendant is suffering from a deficit in rational control can cut both ways—not only reducing perceived culpability but also strengthening the view that such persons cannot be rehabilitated and constitute a greater future risk to society.

Such neurobiological and genetic arguments present an overly deterministic, reductive, and simplistic view of the relationship of genetic variants or differences in size or connectivity of specific brain regions to complex human behaviors occurring in specific individual biographical and sociocultural contexts. Brains do not commit crimes: people do. Of course, genetic and neural variation feeds into our psychological traits, but except for cases of clear pathology, those relationships tend to be indirect, probabilistic, and nonspecific. Moreover, as discussed in chapter ten, our psychological predispositions do not determine our behavior on a moment-to-moment basis; instead, they influence the way our character emerges, but that is a process in which we ourselves play a very active role.

## Some Philosophical Issues

My goal in this book is to present a naturalistic framework for thinking about agency and free will. This depends on or at least involves a reframing of some fundamental philosophical issues, including the nature of causation, time, information, meaning, purpose, and selfhood. It's worth making these positions explicit, especially to better identify the source of possible disagreements.

One of the book's major themes is that the functional organization of a system can have causal power over how things unfold. This seems like a commonplace in our everyday experience: it informs how we design things, how we organize everything from soccer teams to multinational corporations, and how we understand systems of all kinds, from ecologies to economies. We recognize that top-down organization can impose useful constraints that enable functions not achievable by disorganized parts. Why is it then that an idea so fundamental and ubiquitous in our everyday lives is seen as so contentious when applied to living beings?

Part of the resistance, I think, comes from our Western scientific tradition, which, since the time of Newton and Galileo, has focused on bottom-up mechanistic explanations derived from the action of universal physical laws. The staggering successes of physics in predicting ever-finer details of the nature of the fundamental constituents of the universe have bolstered the view that *all the real causes* of what happens are to be found at these lowest levels.

The idea that the organization is also doing causal work at the same time runs into two stubborn problems. First, if the causation underlying how the system evolves from moment to moment can be fully explained by the interactions of the components (bottom-up), there doesn't seem to be any need for higher-order (top-down) causation: we would just have too many explanations of the same thing. Second, if we say that the organization of the whole constrains the parts, we also have to allow that that organization *comes from* the parts. That feels like a circular argument—as if the system would be *the cause of itself*—which is not philosophically tenable.

We discussed, many times, the solution to the first problem. The low-level details and laws of physics do not *fully explain* how the system evolves: not "fully" because indeterminacy leaves open multiple possible paths, and not "explain" because, even if those low-level details allowed you to *predict* how the system evolves from its state "at time *t*," they do not *explain* why it was organized in that way to begin with. The laws of physics answer a *how* question but not the *why* question.

This highlights the crucial nature of causation in living systems: it is extended through time, which in turn provides a resolution of the whole–part circular self-causation problem. The process only looks circular from the perspective of a static point in time, where you have the whole and you have its parts, and it doesn't seem to make sense that intrinsic causation can explain both at the exact same instant. If A causes B and B causes C, it doesn't seem right that C could be causing A at the same time. This leads to the related question: In living beings with nervous systems, how could mental states determine what the physical bits do, when the mental states depend on the arrangement of those bits in the first place?

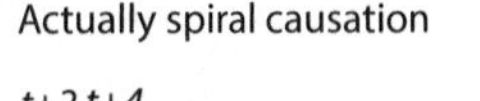

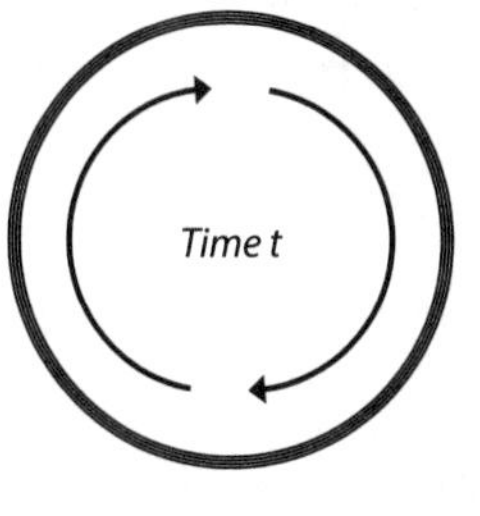


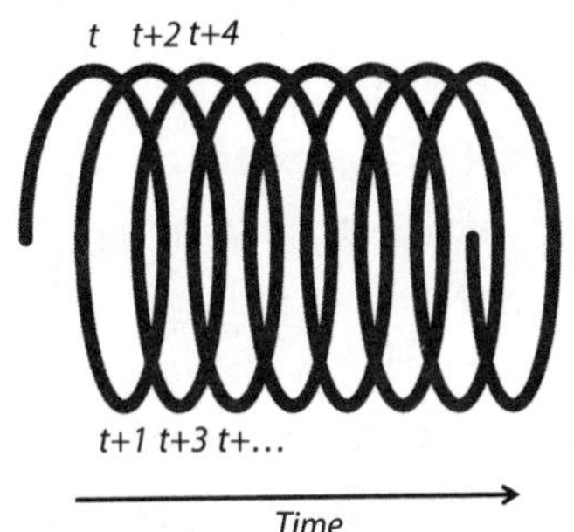


FIGURE 12.1. Spiral causation. A long-standing argument against holistic or top-down causation is that it constitutes a situation of circular causation—where the whole is supposed both to cause the organization of the parts and to be caused by that organization *at the same time*. Taking time into account reveals the true pattern of spiral causation, where interactions up and down the hierarchy are not occurring instantaneously but are spread over time.

But this problem is an illusion of perspective. Causation in the system cannot be understood with an instantaneous perspective. What appears circular is actually a spiral, extended through time (see Figure 12.1). It's like looking at a Slinky (a child's toy, which is basically a big, very flexible metal spring) head-on, when it looks like a perfect circle. But from the side, if you pull on it a bit, you can see it is really a spiral. That side perspective is showing the dimension of time.

Living beings do not cause themselves in an instant, but they do cause themselves through time. That's what being alive entails—continuing to cause yourself. All the components continually regenerate the constraints that keep the whole thing the way it is—*itself*. But this depends on processes that occur over time, not instantaneously. Similarly, top-down mental states do not simultaneously depend on specific lower-level states and determine the same lower-level states in an instant. They are realized in lower-level arrangements, *and then* they influence *subsequent* lower-level arrangements.

Understanding why higher-order patterns have such causal power depends on the notions of purpose, meaning, and value. All these also come with some philosophical baggage. The idea of purpose, in particular, was largely banished from polite scientific discourse because of its connotation of some kind of cosmic teleology. There is indeed no support for the view that evolution itself is purposive—that it reflects *a drive* to create more and more complex life, a drive that could only be ascribed to some godlike entity or to the universe itself. Evolution is a blind, iterative process of variation and selection that allows progressive exploration and discovery of forms with functionalities that tend to favor their own persistence. However, while not purposive itself, evolution does imbue the resulting life-forms with their own purpose. It means that living organisms are genuinely goal directed.

This discussion highlights another type of causation that is also sometimes seen as problematic. Aristotle called it the *final cause*, which doesn't mean the last one but instead the goal toward which some action is aimed. You may put the kettle on *because* you want a cup of tea. You having the cup of tea is seen as the final cause in that scenario; it is a perfectly valid explanation for why you put the kettle on. The problem in that framing is that the cause seems *to follow* the effect, rather than the other way around. How could something in the future really have a causal effect on a physical system in the present?

Again, this framing is not quite right. The cause in this situation is not you having the cup of tea—it's you *wanting* to have the cup of tea. And the reason you want a cup of tea is because you've had them before and you know you enjoy them, especially now that you're thirsty and maybe in need of a little jolt of caffeine. So really it's the accumulation of causes from the past that leads to your current state having causal efficacy in your current decision making. There is thus nothing spooky about goal directedness: it does not violate the normal arrow of causation. All the prior effects of natural selection and learning configure living systems such that they can represent goals and choose actions to achieve them.

This purposiveness also grounds meaning and value. Objects in the world, situations in which an organism finds itself, and the real or predicted outcomes of possible actions all have value for the organism, relative to its ultimate goal of persisting or a host of more proximal goals. The fancy term for this is *normativity*—things mattering. It doesn't exist in nonliving systems: nothing matters to a star or a storm or a carbon atom. Normativity can, however, be designed into artificial systems, which may make some judgments based on what matters to the designer. By analogy, you could say that the simplest living organisms act for natural selection's reasons. But the ability to learn from experience and to engage in real cognitive control granted more complex organisms the capacity to act *for their own reasons.*

This brings us to a recurring concept in this book—meaning—which is another troublesome concept from a scientific perspective. Unlike information, which has a perfectly respectable scientific framework and technological industry built around it, meaning cannot be quantified or localized, either in space or time. It is inherently historical, relational, contextual, qualitative, even subjective. The meaning of the words I am writing now is not in the squiggles on the page or the sounds in your headphones. It's not just in my mind, in the ideas I am trying to capture and convey. And it's not just in how you interpret them, which is based on their context and your own history. The meaning inheres in the relations among all those things, which is spread through space and extended through time. That makes the concept of meaning at best awkward and at worst fatally suspect from a scientific perspective. Yet none of that implies it isn't real.

Meaning clearly has causal power. If I ask you, for example, to raise your right hand or blink your eyes, then the meaning of those words can causally influence your actions. It's not energy or momentum or any physical forces but only meaning that explains these actions. Of course, they require physical processes, including neural patterns flowing from your eyes or ears to other parts of your brain involved in action selection. But those patterns only have causal power by virtue of what they mean.

It has been said that nothing in biology makes sense except in the light of evolution. A more general way to put this is that nothing in biology makes sense except through time. Biology is a historical science because life is a historical process. That view lets us understand the concepts of purpose and meaning and value: they only make sense in light of the selection of living organisms for persistence through time. And the behavior of living organisms only makes sense in light of those concepts.

These views align with what is known as *process philosophy*: the view that *processes*, rather than static isolated objects or *substances*, are the fundamental elements of reality. The idea is that everything we see is really something in flux, changing through time, with some things changing much more rapidly than others. This view recognizes that time is not made up of static instants, in which the state of reality can be frozen and defined in precise detail. Rather, it is made of durations during which everything is in some degree of flux. In the early twentieth century, process philosophy had vocal proponents, such as Alfred North Whitehead and Henri Bergson. It fell out of favor for almost a century but seems to be gathering support again, especially in the philosophy of biology. In particular, it fits with the realization that living organisms maintain themselves *by changing*, as argued by philosopher Anne-Sophie Meincke, among others. They are patterns of self-regenerating processes that are always in flux. The flux is life itself.

The process perspective also jibes with relational views in physics. Carlo Rovelli, for example, argues that objects do not have properties in themselves or really any existence that can be isolated from all their relations to everything else in the universe. Selves are similarly defined by a web of relations: internal relations among all their components and relations with things in the environment, with past experiences and future goals, with biographical memories and personal narratives, and, for humans, with all the other people who ground our own identity.

These views are essentially holistic, even *ecological*: they consider the dynamics of a system in the context of all its relations, both internal and external. For a process as richly complex as life, this feels to me like the right approach. It stands in contrast, however, to the approaches that have dominated biology for decades, which are strongly reductive and mechanistic. There are many reasons for their dominance, some theoretical but a lot simply reflecting methodological limitations. Admittedly, the reductive experimental approach has proven to be extremely powerful and productive, although there is a sense perhaps that it gives a slightly illusory impression of understanding.

The typical approach in trying to figure out how some biological process works is to monitor or manipulate some component while holding everything else constant, thereby allowing you to see what role that component plays in that process. This might mean knocking out a single gene and seeing what effect it has on some phenotype, or stimulating some discrete population of neurons and looking at the effects on behavior, or measuring how some internal proteins change in response to a signal from outside the cell. This design is very powerful, aimed at coping with the complexity of a system by functionally isolating components for study. The problem comes from thinking that the success of those approaches means the system really works like that.

No gene or protein or population of neurons does anything by itself: each is always part of an extended system, and its activities are typically modulated by the context when the organism is behaving in the real world. Finding that some such gene or protein or neuron is involved in one process, under some experimental conditions, does not mean it doesn't also play a role in other, sometimes quite distinct processes. The reductive view of linear pathways with dedicated components is thus an illusion—a product of a forced perspective. It encourages the view of cells or organisms as stimulus–response machines when really they are holistic, proactive, dynamic systems acting in a goal-directed manner; that is, they are agents. They're not pushed around by things in the world, and they're not pushed around by their own parts. This conclusion is reinforced by the dismal record in the pharmaceutical industry of translating supposedly clear scientific results into the clinic, where messy, complex nature typically has the last word.

This too is changing, however. New technologies are enabling researchers to move beyond studying components in isolation to measure instead the behavior of all the genes or proteins in a cell or of tens or hundreds of thousands of neurons at a time—even the entire nervous system in some simple organisms. Coupled with massively increased computational power, access to these kinds of data is prompting a revival of interest in areas like general systems theory, control theory, and cybernetics. These fields had their heyday from the 1950s to the 1970s, based on holistic theories of systems that highlighted the dynamics that emerge with feedback control, for example. They never quite fulfilled their promise, however, mostly because of a lack of data and the computational power to deal with it. That seems to be changing, and we are witnessing a move away from reductive thinking toward an embrace of the reality of living organisms as essentially holistic, integrated systems.

## Transcendence

There is space for one final speculative thought. The view I presented of the evolution of nervous systems is a ruthlessly utilitarian one. Living organisms need to keep their inner processes persisting in the face of an ever-changing environment. They do that initially by regulating their own physiology, but the invention of behavior—the ability to move in the world in a goal-directed fashion—gave organisms the means to extend those regulatory processes through loops with the environment. Simple nervous systems and eventually much more complex brains thus evolved as systems to coordinate action.

All the functions of the brain can be seen through this lens. Perception is all about mapping threats and opportunities and possibilities for action. Cognition is aimed at mapping and knowing the world, understanding causal relations, predicting and anticipating events and the consequences of our own actions to guide effective behavior. Decision making and action selection are directed at prioritizing the optimal goals and actions to support persistence and reproduction. Even consciousness itself may have evolved in the service of better cognitive control and self-regulation. All these neural resources give us the power to optimally guide our behavior over longer timeframes than any other animal.

That's not a very romantic view, however, and I don't want to leave it there, because there is more to our story. In biology, we often try to understand function in terms of adaptation—what some process *was selected for*. But there is another process that is also prevalent throughout evolution, which is known as *exaptation*. This is when a structure that was selected for one function turns out to be useful for some other, often completely novel, purpose. Co-opting feathers to support flight, even though they were probably initially selected for thermoregulation, is a prime example. In many cases, the emergence of one function can offer entry to a new ecological niche where new possibilities abound.

In humans, a kind of cognitive exaptation seems to have occurred. Thinking may have evolved for controlling action. But the expansion of our neural resources and the recursive architecture of our cognitive systems gave us the ability to think about our thoughts. We internalized cognition to such an extent that it became its own world: what cognitive scientists Uta and Chris Frith have called *a world of ideas*. Our minds were set free. We are capable of open-ended, truly creative thought; of imagination; of entertaining fanciful notions and hypothetical futures; of creating art and music and science; and of abstract reasoning that has revealed the deepest laws and principles of the universe.

And we don't do this alone: the true power of human thought comes through collective interaction and cumulative culture. We share and accumulate knowledge and deeper understanding over generations, with young people easily grasping concepts that were literally unthinkable just decades earlier. We have, as individuals and as a species, the power to transcend the immediacies of our own biology. And, though as I write this the global prospects seem gloomy, we have within our reach the possibility of wisdom, of making optimal collective decisions for the long-term survival of our planet and all the wondrous life-forms on it, if we choose to exercise it.

EPILOGUE

# Artificial Agents

A natural question to ask, given the framework I laid out in this book, is whether it can tell us anything useful about the kinds of functional architectures that could underlie true artificial general intelligence or even artificial life. Indeed, one major implication is that *intelligence* and *life* may be inextricably intertwined, in ways that revolve around the concept of agency.

The field of artificial intelligence (AI) has always taken inspiration from neuroscience, starting with the field's founding papers, which suggested that neurons can be thought of as performing logical operations. Taking a cue from that perspective, most of the initial efforts to develop AI focused on tasks requiring abstract, logical reasoning, especially in testing grounds like playing chess or Go, for example—the kinds of things that are hard for most humans. The successes of the field in these arenas are well known.

Recent years have witnessed stunning advances in other areas like image recognition, text prediction, speech recognition, and language translation. These were achieved mainly due to the development and application of *deep learning*, inspired by the massively parallel, multilevel architecture of the cerebral cortex. This approach is tailor-made for learning the statistical regularities in masses and masses of training data. The trained neural networks can then abstract higher-order patterns; for example, recognizing types of objects in images. Or they can predict what patterns will be most likely in new instances of similar data, as in the autocompletion of text messages or the prediction of the three-dimensional structures of proteins.

When trained in the right way, the neural networks can also generate wholly new examples of types of data they have seen before. *Generative models* can be used, for example, to create "a realistic photo image of a horse on the top of Mt. Everest" or a "picture of an ice cream van in the style of van Gogh." And "large language models" can produce what look like very reasonable and cogent passages of text or responses to questions. Indeed, they are capable of having *conversations* that give a strong impression that they truly understand what they are being asked and what they are saying—to the point where some users even attribute sentience to these systems.

However, even the most sophisticated systems can quickly be flummoxed by the right kind of questioning, the kind that presents novel scenarios not represented in the training data that humans can handle quite easily. Thus, if these systems have any kind of "understanding"—based on the abstraction of statistical patterns in an unimaginably vast set of training data—it does not seem to be the kind that humans have.

Indeed, while reaching superhuman performance in many areas, AI has not achieved the same success in things that most humans find easy: moving around in the world, understanding causal relations, or knowing what to do when faced with a novel situation. Notably, these are things that most animals are good at too: they have to be to survive in challenging and dynamic environments.

These limitations reflect the fact that current AI systems are highly specialized: they're trained to do specific tasks on the basis of the patterns in the data they encountered. But when asked to generalize, they often fail, in ways that suggest they did not, in fact, abstract any knowledge of the underlying causal principles at play. They may "know" that when they see X, it is often followed by Y, but they may not know *why* that is: whether it reflects a true causal pattern or merely a statistical regularity, like night following day. They can thus make predictions for familiar types of data but often cannot translate that ability to other types or to novel situations.

Thus, the quest for artificial *general* intelligence has not made the same kind of progress as AI systems aimed at particular tasks. It is precisely that ability to generalize that we recognize as characteristic of natural intelligence. The mark of intelligence in animals is the ability to act appropriately in novel and uncertain environments by applying knowledge and understanding gained from past experience to predict the future, including the outcomes of their own possible actions. Natural intelligence thus manifests in *intelligent behavior*, which is necessarily defined normatively as good or bad, relative to an agent's goals. To paraphrase Forrest Gump, intelligent is as intelligent does.

The other key aspect of natural intelligence is that it is achieved with limited resources. That includes the computational hardware, the energy involved in running it, the amount of experience required to learn useful knowledge, and the time it takes to assess a novel situation and decide what to do. Greater intelligence is the ability not just to arrive at an appropriate solution to a problem but to do so efficiently and quickly. Living organisms do not have the luxury of training on millions of data points, or running a system taking megawatts of power, or spending long periods of time exhaustively computing what to do. It may in fact be precisely those real-world pressures that drive the need and, hence, the ability to abstract general causal principles from limited experience.

Understanding causality can't come from passive observation, because the relevant counterfactuals often do not arise. If X is followed by Y, no matter how regularly, the only way to really know that is a causal relation is to intervene in the system: to prevent X and see if Y still happens. The hypothesis has to be tested. Causal knowledge thus comes from causal intervention in the world. What we see as intelligent behavior is the payoff for that hard work.

The implication is that artificial general intelligence will not arise in systems that only passively receive data. They need to be able to *act back on the world* and see how those data change in response. Such systems may thus have to be embodied in some way: either in physical robotics or in software entities that can act in simulated environments.

Artificial general intelligence may have to be earned through the exercise of agency.

Artificial agents will also need a reason to act one way or another; that is, some kind of master function that can motivate action. In living organisms, that master function is persisting (either individually or through reproduction). All kinds of subsidiary goals can then be scaffolded on top of that master function. Having goals gives actions value: it gives organisms a reason to do one thing versus another or indeed to do anything. It also gives them a reason to learn from experience; specifically, those experiences that are most relevant to their goals and that are most informative about the underlying causal nature of the world.

Living organisms are what philosopher Hans Jonas called *a locus of concern*: things matter to them. And they need the tools to figure out which things matter and in what ways and what they should do about them. As we saw, evolution solved this problem by building a complex architecture of interacting systems for perception, memory, motivation, simulation, selection of goals and plans and actions, reward, reinforcement, learning, and even metacognition. Natural brains thus come with much more structure than current artificial neural networks, and this extended control architecture grounds intelligence in the service of behavior.

For embodied artificial agents to get to the stage where they exhibit general intelligence, they may similarly need an *inbuilt* architecture encompassing those kinds of functionalities. They may even need what we could call *instincts,* including a master function, that give value to states and actions, and possibly some set of preconfigured control policies ("innate" approach or avoidance responses, for example) and learning heuristics that will bias learning in productive ways.

With that kind of scaffold in place, artificial agents could accumulate knowledge in much the way that living organisms do—through motivated interaction with the world, making sense of their environments, learning new associations and causal relations, and abstracting higher-order categories, contingencies, and principles that collectively enable understanding and generalization. Crucially, the *meaning* of these acquired patterns would be grounded by interaction with the world and relative to the agent's own goals and valuation systems. They would mean something, not just to the agent's programmers looking in from the outside but also *to the agents themselves.*

Let me offer one final thought—perhaps the most speculative but also potentially the most fundamental. For artificial systems to have genuine agency—real causal autonomy in the world—there may need to be some indeterminacy in their low-level workings. This could come from their physical components or be programmed in as an added element of randomness or noise. As in living organisms, there may need to be some causal slack at the bottom to allow the agent as a whole to decide—to really *do* anything. Otherwise, no matter how complex, the agent will be pushed around deterministically by its own components. And it will consequently lack the flexibility and creativity needed to drive the kind of exploration of the world and of its own options that will allow it to accumulate causal understanding.

In summary, evolution has given us a roadmap of how to get to intelligence: by building systems that can learn but in a way that is grounded in real experience and causal autonomy from the get-go. You can't build a bunch of algorithms and expect an entity to pop into existence. Instead, you may have to build something with the architecture of an entity and let the algorithms emerge. To get intelligence you may have to build *an intelligence.*

The next question will be whether we should.

# FIGURE CREDITS

1.1 Panels in (c) adapted from K. J. Mitchell, *Innate* (Princeton: Princeton University Press, 2018), 8.

1.2 Human head adapted from C. A. Edwards, A. Kouzani, K. H. Lee, & E. K. Ross (2017), Neurostimulation devices for the treatment of neurologic disorders, *Mayo Clinic Proceedings*, 92(9), 1427–44; mouse adapted from D. Kwon (2018), DBS with nanoparticle-based optogenetics modifies behavior in mice, *The Scientist*, https://www.the-scientist.com/daily-news/dbs-with-nanoparticle-based-optogenetics-modifies-behavior-in-mice-30303.

3.4 Adapted from C. R. Reid & T. Latty (2016), Collective behaviour and swarm intelligence in slime moulds, *FEMS Microbiology Reviews*, *40*(6), 798–806.

4.4 Adapted from C. Dupre & R. Yuste (2017), Non-overlapping neural networks in Hydra vulgaris, *Current Biology*, 27(8), 1085–97.

4.5 Panel A adapted from R. J. Hobson, K. J. Yook, & E. M. Jorgensen (2017), Genetics of neurotransmitter release in *Caenorhabditis elegans*, in *Reference Module in Life Sciences* (Amsterdam: Elsevier); panel B adapted from S. Patil, K. Zhou, & A. Parker (2015), Neural circuits for touch-induced locomotion in *Caenorhabditis elegans*, 2015 International Joint Conference on Neural Networks (IJCNN).

5.2 Reprinted from K. J. Mitchell, *Innate* (Princeton: Princeton University Press, 2018), 127.

5.3 Adapted from P. Cisek (2019), Resynthesizing behavior through phylogenetic refinement, *Attention, Perception, & Psychophysics*, *81*(7), 1165–87.

5.4 Panel A adapted from N. M. Gage & B. J. Baars, *Fundamentals of Cognitive Neuroscience*, 2nd ed. (Cambridge: Academic Press, 2018), 17–52.

5.5 Brain: K. J. Mitchell, *Innate* (Princeton: Princeton University Press, 2018), 130; checker illusion: adapted from Wikipedia user Edward H. Adelson, https://en.wikipedia.org/wiki/Checker_shadow_illusion#/media/File:Checker_shadow_illusion.svg, CC BY-SA 4.0; vehicle illusion: "Logic Optical Illusions," Genius Puzzles, https://gpuzzles.com/optical-illusions/logic/.

6.1 Adapted from P. Cisek (2019), Resynthesizing behavior through phylogenetic refinement, *Attention, Perception, & Psychophysics*, *81*(7), 1165–87.

6.3 Modified from image by OpenClipart-Vectors from Pixabay.

8.1 Modified from K. L. Briggman, H. D. I. Abarbanel, & W. B. Kristan Jr. (2005), Optical imaging of neuronal populations during decision-making, *Science, 307*(5711), 896–901.

8.3 Reprinted, with permission, from P. Haggard (2008), Human volition: Towards a neuroscience of will, *Nature Reviews Neuroscience, 9*(12), 934–46.

8.4 Reprinted from A. Schurger, P. Hu, J. Pak, & A. L. Roskies (2021), What is the readiness potential? *Trends in Cognitive Science, 25*(7), 558–70.

10.2 Reprinted from K. J. Mitchell, *Innate* (Princeton: Princeton University Press, 2018), 127.

11.1 Adapted from A. Nieder (2009), Prefrontal cortex and the evolution of symbolic reference, *Current Opinion in Neurobiology, 19*(1), 99–108.

11.2 Adapted from S. Dehaene, J. Changeux, and L. Naccache (2011), The global neuronal workspace model of conscious access: From neuronal architectures to clinical applications, *Research and Perspectives in the Neurosciences, 18*, 55–84.

# BIBLIOGRAPHY

## Chapter 1—Player One

Carroll, S. (2016). *The big picture: On the origins of life, meaning, and the universe itself.* New York: Dutton.

Deisseroth, K. (2015). Optogenetics: 10 years of microbial opsins in neuroscience. *Nature Neuroscience, 18*(9), 1213–25.

Dennett, D. C. (1984). *Elbow room.* Cambridge, MA: MIT Press.

Greene, Brian (@bgreene). (2013, June 5). Twitter: 9:23 P.M. https://twitter.com/bgreene/status/342376183519916033?lang=en.

Haggard P. (2010, October 12). Neuroscience, free will and determinism: "I'm just a machine." Interview by Tom Chivers for *The Telegraph,* 12. https://www.telegraph.co.uk/news/science/8058541/Neuroscience-free-will-and-determinism-Im-just-a-machine.html.

Harris, S. (2012). *Free will.* New York: Free Press.

Hawking, S., & Mlodinow, L. (2010). *The grand design.* New York: Bantam Books.

Mitchell, K. J. (2018). *Innate: How the wiring of our brains shapes who we are.* Princeton, NJ: Princeton University Press.

Penfield, W. (1961). Activation of the record of human experience. *Annals of the Royal College of Surgeons of England, 29*(2), 77–84.

Schopenhauer, A. (1960). *Essay on the freedom of the will.* New York: Dover.

## Chapter 2—Life Goes On

Deacon, T. W. (2012). *Incomplete nature: How mind emerged from matter.* New York: W. W. Norton.

Juarrero, A. (2002). *Dynamics in action: Intentional behavior as a complex system.* Cambridge, MA: MIT Press.

Lane, N. (2015). *The vital question: Why is life the way it is?* New York: Profile Books.

Martin, W. (2011). Early evolution without a tree of life. *Biology Direct, 6,* 36.

Maturana, H. R., & Varela, F. J. (1980). *Autopoiesis and cognition: The realization of the living,* Vol. 42. Boston Studies in the Philosophy of Science. New York: Springer.

Montévil, M., & Mossio, M. (2015). Biological organisation as closure of constraints. *Journal of Theoretical Biology, 372,* 179–91.

Monty Python. *Dead parrot.* http://montypython.50webs.com/scripts/Series_1/53.htm

Prigogine, I., & Stengers, I. (1984). *Order out of chaos: Man's new dialogue with nature*. London: Heinemann.

Pross, A. (2016). *What is life? How chemistry becomes biology*, 2nd ed. Oxford: Oxford University Press.

Rosen, R. (2005). *Life itself: A comprehensive inquiry into the nature, origin, and fabrication of life*. New York: Columbia University Press.

Schrödinger, E. (1944). *What is life? The physical aspects of the living cell*. Cambridge: Cambridge University Press.

Smith, E., & Morowitz, H. J. (2016) *The origin and nature of life on earth: The emergence of the fourth geosphere*. Cambridge: Cambridge University Press.

Walker, S. I., & Davies, P. C. W. (2013). The algorithmic origins of life. *Journal of the Royal Society Interface, 10*, 20120869.

Woese, C. R. (2002). On the evolution of cells. *Proceedings of the National Academies of Science USA, 99*(13), 8742–47.

Yeats, W. B. (1920). *The Second Coming*. Dublin: Dial.

## Chapter 3—Action!

Bray, D. (2011). *Wetware: A computer in every living cell*. New Haven, CT: Yale University Press.

Brette, R. (2021). Integrative neuroscience of Paramecium, a "swimming neuron." *eNeuro, 8*(3), ENEURO.0018-21.2021.

Dretske, F. (1988). *Explaining behavior: Reasons in a world of causes*. Cambridge, MA: MIT Press.

Gleick, J. (2011). *The information: A history, a theory, a flood*. New York: Pantheon.

Kolchinsky, A., & Wolpert, D. H. (2018). Semantic information, autonomous agency and non-equilibrium statistical physics. *Interface Focus, 8*, 20180041.

Krakauer, D. C., Müller, L., Prohaska, S. J., & Stadler, P. F. (2016). Design specifications for cellular regulation. *Theory in Biosciences, 135*(4), 231–40.

Lyon, P. (2006). The biogenic approach to cognition. *Cognitive Processes, 7*(1), 11–29.

Lyon, P. (2015). The cognitive cell: Bacterial behavior reconsidered. *Frontiers in Microbiology, 6*, 264.

Nicholson, D. J. (2019). Is the cell really a machine? *Journal of Theoretical Biology, 477*, 108–26.

Porter, S. L., Wadhams, G. H., & Armitage, J. P. (2011). Signal processing in complex chemotaxis pathways. *Nature Reviews Microbiology, 9*(3), 153–65.

Potter, H. D., & Mitchell, K. J. (2022). Naturalising agent causation. *Entropy, 24*(4), 472.

Rovelli, C. (2016). Meaning = information + evolution. *arXiv*, 1611.02420v1.

Shannon, C. E. (1948). A mathematical theory of communication. *Bell System Technical Journal, 27*, 379–423.

Sourjik, V., & Wingreen, N. S. (2012). Responding to chemical gradients: Bacterial chemotaxis. *Current Opinion Cell Biology, 24*(2), 262–68.

Von Uexkull, J. (1957). A stroll through the worlds of animals and men. In C. Schiller (Ed. and Trans.), *Instinctive behavior*, 5–80. New York: International Universities Press.

Wang, Y., Chen, C-L., & Iijima, M. (2011). Signaling mechanisms for chemotaxis. *Development, Growth, and Differentiation*, *53*(4), 495–502.

## Chapter 4—Life Gets Complicated

Arendt, D., Tosches, M. A., & Marlow, H. (2016). From nerve net to nerve ring, nerve cord and brain—evolution of the nervous system. *Nature Reviews Neuroscience*, *17*(1), 61–72.

Brunet, T., & King, N. (2017). The origin of animal multicellularity and cell differentiation. *Developmental Cell*, *43*(2), 124–40.

Dupre, C., & Yuste, R. (2017). Non-overlapping neural networks in Hydra vulgaris. *Current Biology*, *27*(8), 1085–97.

Folse, H. J. III, & Roughgarden, J. (2010). What is an individual organism? A multilevel selection perspective. *Quarterly Review of Biology*, *85*(4), 447–72.

Keijzer, F. (2015). Moving and sensing without input and output: Early nervous systems and the origins of the animal sensorimotor organization. *Biology & Philosophy*, *30*(3), 311–31.

Kundert, P., & Shaulsky, G. (2019). Cellular allorecognition and its roles in Dictyostelium development and social evolution. *International Journal of Developmental Biology*, *63*(8-9-10), 383–93.

Lane, N. (2011). Energetics and genetics across the prokaryote-eukaryote divide. *Biology Direct*, *6*, 35.

Martin, W. F., Garg, S., & Zimorski, V. (2015). Endosymbiotic theories for eukaryote origin. *Philosophical Transactions of the Royal Society London B: Biological Sciences*, *370*(1678), 20140330.

Michod, R. E. (2007). Evolution of individuality during the transition from unicellular to multicellular life. *Proceedings of the National Academies of Science USA*, *104*(Suppl 1), 8613–18.

Niklas, K. J., & Newman, S. A. (2013). The origins of multicellular organisms. *Evolution and Development*, *15*(1), 41–52.

Rahmani, A., & Chew, Y. L. (2021). Investigating the molecular mechanisms of learning and memory using Caenorhabditis elegans. *Journal of Neurochemistry*, *159*(3), 417–51.

Sagan, L. (née Margulis). (1967). On the origin of mitosing cells. *Journal of Theoretical Biology*, *14*(3), 225–74.

Sasakura, H., & Mori, I. (2013). Behavioral plasticity, learning, and memory in C. elegans. *Current Opinion in Neurobiology*, *23*(1), 92–99.

Sterling, P., & Laughlin, S. (2015). *Principles of neural design*. Cambridge, MA: MIT Press.

Stern, S., Kirst, C., & Bargmann, C. I. (2017). Neuromodulatory control of long-term behavioral patterns and individuality across development. *Cell*, *171*(7), 1649–62.e10.

Szymanski, J. R., & Yuste, R. (2019). Mapping the whole-body muscle activity of Hydra vulgaris. *Current Biology*, *29*(11), 1807–17.e3.

## Chapter 5—The Perceiving Self


Arendt, D. (2003). Evolution of eyes and photoreceptor cell types. *International Journal of Developmental Biology, 47*(7–8), 563–71.

Buzsáki, G. (2019). *The brain from inside out.* Oxford: Oxford University Press.

Cisek, P. (2019). Resynthesizing behavior through phylogenetic refinement. *Attention, Perception, & Psychophysics, 81*(7), 2265–87.

Clark, A. (2016). *Surfing uncertainty: Prediction, action, and the embodied mind.* Oxford: Oxford University Press.

Damasio, A. (2018). *The strange order of things: Life, feeling, and the making of cultures.* New York: Pantheon Books.

Dretske, F. (1988). *Explaining behavior: Reasons in a world of causes.* Cambridge, MA: MIT Press.

Godfrey-Smith, P. (2020). *Metazoa: Animal life and the birth of the mind.* New York: Farrar, Straus and Giroux.

Grillner, S., & Robertson, B. (2016). The basal ganglia over 500 million years. *Current Biology, 26*(20), R1088–R1100.

Jékely, G., Godfrey-Smith, P., & Keijzer F. (2021). Reafference and the origin of the self in early nervous system evolution. *Philosophical Transactions of the Royal Society London B: Biological Sciences, 376*(1821), 20190764.

Kaas, J. H. (2013). The evolution of brains from early mammals to humans. *Interdisciplinary Review of Cognitive Science, 4*(1), 33–45.

Krubitzer, L. (2009). In search of a unifying theory of complex brain evolution. *Annals of the New York Academy of Science, 1156*, 44–67.

MacIver, M. A., & Finlay, B. L. (2022). The neuroecology of the water-to-land transition and the evolution of the vertebrate brain. *Philosophical Transactions of the Royal Society London B: Biological Sciences, 377*(1844), 20200523.

Millikan, R. (1989). Biosemantics. In Brian P. McLaughlin & Ansgar Beckerman (Eds.), *Journal of Philosophy*, 281–97. Oxford: Oxford University Press.

Nilsson, D. E. (2009). The evolution of eyes and visually guided behaviour. *Philosophical Transactions of the Royal Society London B: Biological Sciences, 364*(1531), 2833–47.

Pezzulo, G., & Castelfranchi, C. (2007). The symbol detachment problem. *Cognitive Processes, 8*(2), 115–31.

Pezzulo, G., & Cisek, P. (2016). Navigating the affordance landscape: Feedback control as a process model of behavior and cognition. *Trends in Cognitive Science, 20*(6), 414–24.

Puelles, L., & Rubenstein, J. L. (2015). A new scenario of hypothalamic organization: Rationale of new hypotheses introduced in the updated prosomeric model. *Frontiers in Neuroanatomy, 9*, 27.

Rosa, M. G., & Tweedale, R. (2005). Brain maps, great and small: Lessons from comparative studies of primate visual cortical organization. *Philosophical Transactions of the Royal Society London B: Biological Sciences, 360*(1456), 665–91.

Shea, N. (2018). *Representation in cognitive neuroscience.* Oxford: Oxford University Press.

Teufel, C., & Fletcher, P. C. (2020). Forms of prediction in the nervous system. *Nature Reviews Neuroscience, 21*(4), 231–42.

## Chapter 6—Choosing

Bach, D. R., & Dayan, P. (2017). Algorithms for survival: A comparative perspective on emotions. *Nature Reviews Neuroscience, 18*(5), 311–19.

Buzsáki, G. (2019). *The brain from inside out.* Oxford: Oxford University Press.

Cisek, P. (2019). Resynthesizing behavior through phylogenetic refinement. *Attention, Perception, & Psychophysics, 81*(7), 2265–87.

Douglas, R. J., & Martin, K. A. (2012). Behavioral architecture of the cortical sheet. *Current Biology, 22*(24), R1033-8.

Dudman, J. T., & Krakauer, J. W. (2016). The basal ganglia: From motor commands to the control of vigor. *Current Opinion in Neurobiology, 37*, 158–66.

Graziano, M. S. A. (2016), Ethological action maps: A paradigm shift for the motor cortex. *Trends in Cognitive Science, 20*(2), 121–32.

Grillner, S., & Robertson, B. (2016). The basal ganglia over 500 million years. *Current Biology, 26*(20), R1088–R1100.

Holtmaat, A., & Caroni, P. (2016). Functional and structural underpinnings of neuronal assembly formation in learning. *Nature Neuroscience, 19*(12), 1553–62.

Hwang, E. J. (2013). The basal ganglia, the ideal machinery for the cost-benefit analysis of action plans. *Frontiers in Neural Circuits, 7*, 121.

Isa, T., Marquez-Legorretas E., Grillners S., & Scott, E. K. (2021). The tectum/superior colliculus as the vertebrate solution for spatial sensory integration and action. *Current Biology, 31*(11), R741–R762.

LeDoux, J. (2012). Rethinking the emotional brain. *Neuron, 73*(4), 653–76.

Morris, A., Phillips, J., Huang, K., & Cushman, F. (2021). Generating options and choosing between them depend on distinct forms of value representation. *Psychological Science, 32*(11), 1731–46.

Pezzulo, G., & Cisek, P. (2016). Navigating the affordance landscape: Feedback control as a process model of behavior and cognition. *Trends in Cognitive Science, 20*(6), 414–24.

Redish, A. D. (2013). *The mind within the brain: How we make decisions and how those decisions go wrong.* Oxford: Oxford University Press.

Sherman, S. M., & Usrey, W. M. (2021). Cortical control of behavior and attention from an evolutionary perspective. *Neuron, 109*(19), 3048–54.

Verschure, P. F., Pennartz, C. M., & Pezzulo, G. (2014). The why, what, where, when and how of goal-directed choice: Neuronal and computational principles. *Philosophical Transactions of the Royal Society London B: Biological Sciences, 369*(1655), 20130483.

## Chapter 7—The Future Is Not Written

Bateson, G. (1971). The cybernetics of "self": A theory of alcoholism. *Psychiatry, 34*(1), 1–18.

Batterman, R. W. (2011). Emergence, singularities, and symmetry breaking. *Foundations of Physics, 41*(6), 1031–50.

Bell, J. S. (1964). On the Einstein Podolsky Rosen paradox. *Physics Physique Fizika, 1*(3), 195–200.

Carroll, S. (2016). *The big picture: On the origins of life, meaning, and the universe itself.* New York: Dutton.

Carroll, S. (2021). *Consciousness and the laws of physics* [preprint]. http://philsci-archive.pitt.edu/19311/.

Del Santo, F., & Gisin, N. (2019). Physics without determinism: Alternative interpretations of classical physics. *Physical Review A,* 100.

Dennett, D. C. (1984). *Elbow room.* Cambridge, MA: MIT Press.

Dennett, D. C. (2003). *Freedom evolves.* New York: Viking Press.

Dennett, D. C., & Caruso, G. D. (2021). *Just deserts: Debating free will.* Cambridge: Polity.

Dretske, F. (1988). *Explaining behavior: Reasons in a world of causes.* Cambridge, MA: MIT Press.

Ellis, G. F. R. (2008). On the nature of causation in complex systems. *Transactions of the Royal Society of South Africa, 63*(1), 69–84.

Ellis, G. F. R. (2016). *How can physics underlie the mind? Top-down causation in the human context.* Berlin: Springer.

Everett, H. (1957). Relative state formulation of quantum mechanics. *Reviews of Modern Physics,* 29(3), 454–62.

Falcon, A. Aristotle on causality. (2022, Spring). In Edward N. Zalta (Ed.), *The Stanford encyclopedia of philosophy.* https://plato.stanford.edu/archives/spr2022/entries/aristotle-causality/.

Feynman, Richard. https://en.wikiquote.org/wiki/Talk:Richard_Feynman#%22If_you_think_you_understand_quantum_mechanics,_you_don't_understand_quantum_mechanics.%22.

Gisin, N. (2021). Indeterminism in physics and intuitionistic mathematics. *Synthese, 199,* 13345–71.

Gleick, J. (1996). *Chaos.* London: Vintage.

Gordus, A., Pokala, N., Levy, S., Flavell, S. W., & Bargmann, C. I. (2015). Feedback from network states generates variability in a probabilistic olfactory circuit. *Cell, 161*(2), 215–27.

Greenblatt, S. (2011). *The swerve: How the world became modern.* New York: W. W. Norton.

Hesse, J., and Gross, T. (2014). Self-organized criticality as a fundamental property of neural systems. *Frontiers in System Neuroscience, 8,* 166.

Laplace, P. S. (1902). *A philosophical essay on probabilities.* New York: Wiley & Sons.

Laughlin, R. B., & Pines, D. (2000). The theory of everything. *Proceedings of the National Academy of Sciences USA,* 97(1), 28–31.

Marx, K. (1902). The difference between the Democritean and Epicurean philosophy of nature. In *Marx-Engels Collected Works, Vol. 1.* Moscow: Progress Publishers.

Ross, W. D. (Ed.) (1924). *Aristotle's Metaphysics, Vol. 2.* Oxford: Oxford University Press.

Rovelli, C. (2021). *Helgoland.* London: Allen Lane.

Saunders, T. J. (1984). Free will and the atomic swerve in Lucretius. *Symbolae Osloenses,* 59(1), 37–59.

Sedley, D. (1983). Epicurus' refutation of determinism. *SUZETESIS,* 11–51.

Smolin, L., & Verde, C. (2021). The quantum mechanics of the present. *arXiv,* 2104.09945.

Steward, H. (2012). *A metaphysics for freedom.* Oxford: Oxford University Press.

Tse, P., U. (2013). *The neural basis of free will: Criterial causation*. Cambridge, MA: MIT Press.
Wilczek, F. (2021). *Fundamentals: Ten keys to reality*. London: Penguin.

## Chapter 8—Harnessing Indeterminacy

Aston-Jones, G., & Cohen, J. D. (2005). An integrative theory of locus coeruleus-norepinephrine function: Adaptive gain and optimal performance. *Annual Review of Neuroscience, 28*, 403–50.

Braun, H. A. (2021). Stochasticity versus determinacy in neurobiology: From ion channels to the question of "free will." *Frontiers in System Neuroscience, 26*(15), 629436.

Brembs, B. (2011). Towards a scientific concept of free will as a biological trait: Spontaneous actions and decision-making in invertebrates. *Proceedings of the Royal Society B: Biological Sciences, 278*(1707), 930–39.

Brembs, B. (2021). The brain as a dynamically active organ. *Biochemical and Biophysical Research Communications, 564*, 55–69.

Briggman, K. L., Abarbanel, H. D., & Kristan, W. B. Jr. (2005). Optical imaging of neuronal populations during decision-making. *Science, 307*(5711), 896–901.

Carandini, M. (2004). Amplification of trial-to-trial response variability by neurons in visual cortex. *PLoS Biology 2*, E264.

Deco, G., & Rolls, E. T. (2006). Decision-making and Weber's law: A neurophysiological model. *European Journal of Neuroscience, 24*(3), 901–16.

Dennett, D. C. (1984). I could not have done otherwise—so what? *Journal of Philosophy, 81*(10), 553–65.

Domenici, P., Booth, D., Blagburn, J. M., & Bacon, J. P. (2008). Cockroaches keep predators guessing by using preferred escape trajectories. *Current Biology, 18*(22), 1792–96.

Doyle, R. O. (2016). The two-stage model to the problem of free will: How behavioral freedom in lower animals has evolved to become free will in humans and higher animals. In A. Suarez & P. Adams (Eds.), *Is science compatible with free will? Exploring free will and consciousness in the light of quantum physics and neuroscience*. New York: Springer Science+Business Media.

Ebitz, R. B., & Hayden, B. Y. (2021). The population doctrine in cognitive neuroscience. *Neuron, 109*(19), 3055–68.

Faisal, A. A., Selen, L. P., & Wolpert, D. M. (2008). Noise in the nervous system. *Nature Reviews Neuroscience, 9*(4), 292–303.

Fee, M. S., & Goldberg, J. H. (2011). A hypothesis for basal ganglia-dependent reinforcement learning in the songbird. *Neuroscience, 198*, 152–70.

Frankel, N. W., Pontius, W., Dufour, Y. S., Long, J., Hernandez-Nunez, L., & Emonet, T. (2014). Adaptability of non-genetic diversity in bacterial chemotaxis. *Elife, 3*, e03526.

Glimcher, P. W. (2005). Indeterminacy in brain and behavior. *Annual Review of Psychology, 56*, 25–56.

Gold, J. I., & Shadlen, M. N. (2007). The neural basis of decision making. *Annual Review of Neuroscience, 30*, 535–74.

Heisenberg, M. (2009). Is free will an illusion? *Nature, 459*(7244), 164–65.

Hoel, E. (2021). The overfitted brain: Dreams evolved to assist generalization. *Patterns, 2*(5), 100244.

James, W. (1884). The dilemma of determinism. *Unitarian Review, 32,* 1993. Reprinted in (1956) *The will to believe,* 145. New York: Dover.

Kaplan, H. S., & Zimmer, M. (2020). Brain-wide representations of ongoing behavior: A universal principle? *Current Opinion in Neurobiology, 64,* 60–69.

Libet, B., Gleason, C. A., Wright, E. W., & Pearl, D. K. (1983). Time of conscious intention to act in relation to onset of cerebral activity (readiness-potential): The unconscious initiation of a freely voluntary act. *Brain, 106*(Pt. 3), 623–42.

London, M., Roth, A., Beeren, L., Häusser, M., & Latham, P. E. (2010). Sensitivity to perturbations in vivo implies high noise and suggests rate coding in cortex. *Nature, 466*(7302), 123–27.

Mainen, Z. F., & Sejnowski, T. J. (1995). Reliability of spike timing in neocortical neurons. *Science, 268*(5216), 1503–6.

Maoz, U., Yaffe, G., Koch, C., & Mudrik, L. (2019). Neural precursors of decisions that matter—An ERP study of deliberate and arbitrary choice. *Elife, 8,* e39787.

Musall, S., Kaufman, M. T., Juavinett, A. L., Gluf, S., & Churchland, A. K. (2019). Single-trial neural dynamics are dominated by richly varied movements. *Nature Neuroscience, 22*(10), 1677–86.

Nash, J. F. (1950). Equilibrium points in N-person games. *Proceedings of the National Academies of Science USA, 36*(1), 48–49.

Noble, R., & Noble, D. (2108). Harnessing stochasticity: How do organisms make choices? *Chaos, 28*(10), 106309.

Nolte, M., Reimann, M. W., King, J. G., Markram, H., & Muller, E. B. 2019. Cortical reliability amid noise and chaos. *Nature Communications, 10*(1), 3792.

Poe, G. R., Foote, S., Eschenko, O., Johansen, J. P., Bouret, S., Aston-Jones, G., Harley, C. W., et al. (2020). Locus coeruleus: A new look at the blue spot. *Nature Reviews Neuroscience, 21*(11), 644–59.

Schurger, A., Hu. P., Pak, J., & Roskies, A. L. (2021). What is the readiness potential? *Trends in Cognitive Science, 25*(7), 558–70.

Shadlen, M. N., & Newsome, W. T. (1994). Noise, neural codes and cortical organization. *Current Opinion in Neurobiology, 4*(4), 569–79.

Shew, W. L., & Plenz, D. (2013). The functional benefits of criticality in the cortex. *The Neuroscientist, 19*(1), 88–100.

Simonton, D. K. (2013). Creative thoughts as acts of free will: A two-stage formal integration. *Review of General Psychology, 17*(4), 374–78.

Steinmetz, N. A., Zatka-Haas, P., Carandini, M., & Harris, K. D. (2019). Distributed coding of choice, action and engagement across the mouse brain. *Nature, 576*(7786), 266–73.

Sterling, P., & Laughlin, S. (2015). *Principles of neural design.* Cambridge, MA: MIT Press.

Urai, A. E., Doiron, B., Leifer, A. M., & Churchland, A. K. (2022). Large-scale neural recordings call for new insights to link brain and behavior. *Nature Neuroscience, 25*(1), 11–19.

Von Neumann, J. (1956). Probabilistic logics and synthesis of reliable organisms from unreliable components. In C. E. Shannon and J. McCarthy (Eds.), *Automata studies,* 43–98. Annals of Mathematical Studies No. 34, Princeton, NJ: Princeton University Press.

Von Neumann, J., & Morgenstern, O. (1944). *Theory of games and economic behavior.* Princeton, NJ: Princeton University Press.

## Chapter 9—Meaning

Bechtel, W., & Bich, L. (2021). Grounding cognition: Heterarchical control mechanisms in biology. *Philosophical Transactions of the Royal Society B, 376*, 20190751.

Binder, J. R., & Desai, R. H. (2011). The neurobiology of semantic memory. *Trends in Cognitive Science, 15*(11), 527–36.

Brembs, B. (2021). The brain as a dynamically active organ. *Biochemical and Biophysical Research Communications, 564*, 55–69.

Buzsáki, G. (2010). Neural syntax: Cell assemblies, synapsembles, and readers. *Neuron, 68*(3), 362–85.

Buzsáki, G. (2019). *The brain from inside out.* Oxford: Oxford University Press.

Cisek, P. (2019). Resynthesizing behavior through phylogenetic refinement. *Attention, Perception, & Psychophysics, 81*(7), 2265–87.

Correia, J., Formisano, E., Valente, G., Hausfeld, L., Jansma, B., & Bonte, M. (2014). Brain-based translation: fMRI decoding of spoken words in bilinguals reveals language-independent semantic representations in anterior temporal lobe. *Journal of Neuroscience, 34*(1), 332–38.

Crick, Francis. (1994). *The astonishing hypothesis: The scientific search for the soul.* New York: Scribner.

Deacon, Terrence W. (2012). *Incomplete nature: How mind emerged from matter.* New York: W. W. Norton.

Deco, G., & Rolls, E. T. (2006). Decision-making and Weber's law: A neurophysiological model. *European Journal of Neuroscience, 24*(3), 901–16.

Dohmatob, E., Dumas, G., & Bzdok, D. (2020). Dark control: The default mode network as a reinforcement learning agent. *Human Brain Mapping, 41*(12), 3318–41.

Dretske, F. (1988). *Explaining behavior: Reasons in a world of causes.* Cambridge, MA: MIT Press.

Ebitz, R. B., & Hayden, B. Y. (2021). The population doctrine in cognitive neuroscience. *Neuron, 109*(19), 3055–68.

Ellis, G. F. R. (2016). *How can physics underlie the mind? Top-down causation in the human context.* Berlin: Springer.

Harnad, S. (1990). The symbol grounding problem. *Physica D, 42*, 335–46.

Huth, A. G., Nishimotos S., Vu, A. T., & Gallant, J. L. (2012). A continuous semantic space describes the representation of thousands of object and action categories across the human brain. *Neuron, 76*(6), 1210–24.

Jékely, G., Godfrey-Smith, P., & Keijzer, F. (2021). Reafference and the origin of the self in early nervous system evolution. *Philosophical Transactions of the Royal Society London B: Biological Sciences, 376*(1821), 20190764.

Juarrero, A. (2002). *Dynamics in action: Intentional behavior as a complex system.* Cambridge, MA: MIT Press.

Juarrero Roqué, A. (1985). Self-organization: Kant's concept of teleology and modern chemistry. *Review of Metaphysics, 39*(1), 107–35.

Kant, I. (1790). *The critique of judgement* (J. C. Meredith, trans.). Oxford: Clarendon Press, 1980.

Kauffman, S. (2013). What is life, and can we create it? *BioScience, 63*(8), 609–10.
Kriegeskorte, N., Mur, M., Ruff, D. A., Kiani, R., Bodurka, J., Esteky, H., Tanaka, K., & Bandettini, P. A. (2008). Matching categorical object representations in inferior temporal cortex of man and monkey. *Neuron, 60*(6), 1126–41.
McCulloch, W. S., & Pitts, W. (1943). A logical calculus of the ideas immanent in nervous activity. *Bulletin of Mathematical Biophysics, 5*, 115–33.
Millikan, R. (1989). Biosemantics. In Brian P. McLaughlin & Ansgar Beckerman (Eds.), *Journal of Philosophy*, 281–97. Oxford: Oxford University Press.
Mitchell, K. J. (2018). Does neuroscience leave room for free will? *Trends in Neuroscience, 41*(9), 573–76.
Nakai, T., & Nishimoto, S. (2021). Preserved representations and decodability of diverse cognitive functions across the cortex, cerebellum, and subcortex. *bioRxiv*, 12.09.471939.
Pattee, H. H. (1982). Cell psychology: An evolutionary approach to the symbol-matter problem. *Cognition and Brain Theory*, 5(4), 325–41.
Peirce, C. S. (1978). *The philosophy of Peirce: Selected writings*. New York: AMS Press.
Pezzulo, G., & Castelfranchi, C. (2007). The symbol detachment problem. *Cognitive Processes, 8*(2), 115–31.
Popham, S. F., Huth, A. G., Bilenko, N. Y., Deniz, F., Gao, J. S., Nunez-Elizalde, A. O., & Gallant, J. L. (2021). Visual and linguistic semantic representations are aligned at the border of human visual cortex. *Nature Neuroscience, 24*(11), 1628–36.
Potter, H. D., & Mitchell, K. J. (2022). Naturalising agent causation. *Entropy, 24*(4), 472.
Raichle, M. E. (2010). Two views of brain function. *Trends in Cognitive Science, 14*(4), 180–90.
Saxena, S., & Cunningham, J. P. (2019). Towards the neural population doctrine. *Current Opinion in Neurobiology, 55*, 103–11.
Semedo, J. D., Gokcen, E., Machens, C. K., Kohn, A., & Yu, B. M. (2020). Statistical methods for dissecting interactions between brain areas. *Current Opinion in Neurobiology, 65*, 59–69.
Shadlen, M. N., & Newsome, W. T. (1994). Noise, neural codes and cortical organization. *Current Opinion in Neurobiology, 4*(4), 569–79.
Shea, N. (2018). *Representation in cognitive neuroscience*. Oxford: Oxford University Press.
Tolman, E. C. (1948). Cognitive maps in rats and men. *Psychological Review, 55*, 189–208.
Tse, P. U. (2013). *The neural basis of free will: Criterial causation*. Cambridge, MA: MIT Press.
Walker, S. I. (2014). Top-down causation and the rise of information in the emergence of life. *Information, 5*(3), 424–39.
Whittington, J. C. R., Muller, T. H., Mark, S., Chen, G., Barry, C., Burgess, N., & Behrens, T. E. J. (2020). The Tolman-Eichenbaum machine: Unifying space and relational memory through generalization in the hippocampal formation. *Cell, 183*(5), 1249–63.e23.

## Chapter 10—Becoming Ourselves

Aristotle. (2009). *The Nicomachean Ethics* (W. D. Ross, trans.). Oxford: Oxford University Press.
Bandura, A. (2001). Social cognitive theory: An agentic perspective. *Annual Review of Psychology*, 52 (2001), 1–26.

Banicki, K. (2017). The character–personality distinction: An historical, conceptual, and functional investigation. *Theory and Psychology*, 27(1), 50–68.

Churchland, P. S. (2011). *Braintrust: What neuroscience tells us about morality.* Princeton, NJ: Princeton University Press.

Cicero. (1913). *De Officiis* (Walter Miller, trans.). Cambridge, MA: Harvard University Press.

Cloninger, C. R., Svrakic, D. M., & Przybeck, T. R. (1993). A psychobiological model of temperament and character. *Archives of General Psychiatry, 50*(12), 975–90.

Critchlow, H. (2019). *The science of fate: Why your future is more predictable than you think.* London: Hodder & Stoughton.

Dayan, P. (2012). Twenty-five lessons from computational neuromodulation. *Neuron, 76*, 240–56.

DeYoung, C. G. (2010). Personality neuroscience and the biology of traits. *Social and Personality Psychology Compass*, 4(12), 1165–80.

Dolan, R. J., & Dayan, P. (2013). Goals and habits in the brain. *Neuron, 80*(2), 312–25.

Gill, C. (1998). The question of character-development: Plutarch and Tacitus. *Classical Quarterly*, 33(2), 469–87.

Haidt, J. (2003). The moral emotions. In R. J. Davidson, K. R. Scherer, & H. H. Goldsmith (Eds.), *Handbook of affective sciences*, 852–70. Oxford: Oxford University Press.

Harris, J. R. (1998). *The nurture assumption: Why children turn out the way they do.* New York: Free Press.

Harris, S. (2012). *Free will.* New York: Free Press.

Hiesinger, P. R. (2021). *The self-assembling brain: How neural networks grow smarter.* Princeton, NJ: Princeton University Press.

Hill, P. L., Edmonds, G. W., & Jackson, J. J. (2019). Pathways linking childhood personality to later life outcomes. *Child Development Perspectives*, 13(2), 116–20.

Kahneman, D. (2013). *Thinking, fast and slow.* New York: Farrar, Straus and Giroux.

Konch, M., & Panda, R. K. (2019). Aristotle on habit and moral character formation. *International Journal of Ethics Education*, 4, 31–41.

McAdams, D. P., & Pals, J. L. (2006). A new Big Five: Fundamental principles for an integrative science of personality. *American Psychologist, 61*, 204–17.

Meincke, A. S. (2019). Human persons—A process view. In Jörg Ulrich Noller (Ed.), *Was sind und wie existieren Personen?* 53–76. Münster: Brill.

Mitchell, K. J. (2018). *Innate: How the wiring of our brains shapes who we are.* Princeton, NJ: Princeton University Press.

Nettle, D. (2009). *Personality: What makes you the way you are.* Oxford: Oxford University Press.

Pinker, S. (2002). *The blank slate: The modern denial of human nature.* New York: Viking.

Polderman, T. J. C., Benyamin, B., de Leeuw, C. A., Sullivan, P. F., van Bochoven, A., Visscher, P. M., & Posthuma, D. (2015). Meta-analysis of the heritability of human traits based on fifty years of twin studies. *Nature Genetics*, 47, 702–9.

Poldrack, R. A. (2021). *Hard to break: Why our brains make habits stick.* Princeton, NJ: Princeton University Press.

Raihani, N. (2021). *The social instinct: How cooperation shaped the world.* London: Jonathan Cape.

Scarr, S., & McCartney, K. (1983), How people make their own environments: A theory of genotype-environment effects. *Child Development*, 54, 424–35.

Schopenhauer, A. (1960). *Essay on the freedom of the will.* New York: Dover.

Sherman, N. (1985). Character, planning, and choice in Aristotle. *Review of Metaphysics, 39*(1), 83–106.

Tangney, J. P., Stuewig, J., & Mashek, D. J. (2007). Moral emotions and moral behavior. *Annual Review of Psychology, 58*, 345–72.

Webber, J. (2006). Sartre's theory of character. *European Journal of Philosophy, 14*(1), 94–116.

Willoughby, E. A., Love, A. C., McGue, M., Iacono, W. G., Quigley, J., and Lee, J. J. (2019). Free will, determinism, and intuitive judgments about the heritability of behavior. *Behavioral Genetics, 49*(2), 136–53.

## Chapter 11—Thinking about Thinking

Badre, D., & Nee, D. E. (2018). Frontal cortex and the hierarchical control of behavior. *Trends in Cognitive Science, 22*(2), 170–88.

Bargh, J. A., Gollwitzer, P. M., Lee-Chai, A., Barndollar, K., & Trötschel, R. (2001). The automated will: Nonconscious activation and pursuit of behavioral goals. *Journal of Personal and Social Psychology, 81*(6), 1014–27.

Benjamin, D. J., Laibson, D., Mischel, W., Peake, P. K., Shoda, Y., Wellsjo, A. S., & Wilson, N. L. (2020). Predicting mid-life capital formation with pre-school delay of gratification and life-course measures of self-regulation. *Journal of Economic Behavior and Organization, 179*, 743–56.

Brown, R., Lau, H., & LeDoux, J. E. (2019). Understanding the higher-order approach to consciousness. *Trends in Cognitive Science, 23*(9), 754–68.

Calvin, W. H. (2004). *A brief history of the mind.* Oxford: Oxford University Press.

Dehaene, S. (2017). What is consciousness, and could machines have it? *Science, 358*, 486–92.

Dehaene, S., & Changeux, J. P. (2011). Experimental and theoretical approaches to conscious processing. *Neuron, 70*(2), 200–27.

Dennett, D. C. (2017). *From bacteria to Bach and back: The evolution of minds.* New York: W. W. Norton.

Eisenberg, I. W., Bissett, P. G., Zeynep Enkavi, A., Li, J., MacKinnon, D. P., Marsch, L. A., & Poldrack, R. A. (2019). Uncovering the structure of self-regulation through data-driven ontology discovery. *Nature Communications, 10*(1), 2319.

Fine, J. M., & Hayden, B. Y. (2022). The whole prefrontal cortex is premotor cortex. *Philosophical Transactions of the Royal Society London B: Biological Sciences, 377*(1844), 20200524.

Fleming, S. M. (2021). *Know thyself: The science of self-awareness.* New York: Basic Books.

Freeman, W. J. (1999). *How brains make up their minds.* London: Weidenfeld and Nicolson.

Friedman, N. P., & Miyake, A. (2017). Unity and diversity of executive functions: Individual differences as a window on cognitive structure. *Cortex, 86*, 186–204.

Frith, C. (2007). *Making up the mind: How the brain creates our mental world.* Oxford: Blackwell.

Frith, U., & Frith C. (2010). The social brain: Allowing humans to boldly go where no other species has been. *Philosophical Transactions of the Royal Society London B: Biological Sciences, 365*(1537), 165–76.

Gazzaniga, M. S. (2013). Shifting gears: Seeking new approaches for mind/brain mechanisms. *Annual Review of Psychology, 64,* 1–20.

Goldberg, E. (2011). *The executive brain: Frontal lobes and the civilized mind.* Oxford: Oxford University Press.

Harris, S. (2012). *Free will.* New York: Free Press.

Hofstadter, D. R. (1982). Who shoves whom around inside the careenium? Or what is the meaning of the word "I"? *Synthese, 53,* 189–218.

Hoftstadter, D. R. (2007). *I am a strange loop.* New York: Basic Books.

Hommel, B., Chapman, C. S., Cisek, P., Neyedli, H. F., Song, J. H., & Welsh, T. N. (2019). No one knows what attention is. *Attention, Perception, & Psychophysics, 81*(7), 2288–303.

Kaas, J. H. (2013).The evolution of brains from early mammals to humans. *Interdisciplinary Review of Cognitive Science, 4*(1), 33–45.

Kahneman, D. (2013). *Thinking, fast and slow.* New York: Farrar, Straus and Giroux.

Mattar, M. G., & Lengyel, M. (2022). Planning in the brain. *Neuron, 110*(6), 914–34.

Miller, E. K., & Cohen, J. D. (2001). An integrative theory of prefrontal cortex function. *Annual Review of Neuroscience, 24,* 167–202.

Mischel, W., Shoda, Y., & Rodriguez, M. I. (1989). Delay of gratification in children. *Science, 244*(4907), 933–38.

Miyake, A., Friedman, N. P., Emerson, M. J., Witzki, A. H., Howerter, A., & Wager, T. D. (2000). The unity and diversity of executive functions and their contributions to complex "frontal lobe" tasks: A latent variable analysis. *Cognitive Psychology, 41*(1), 49–100.

Open Science Collaboration. (2015). Estimating the reproducibility of psychological science. *Science, 349*(6251), aac4716.

Preuss, T. M., & Wise, S. P. (2022). Evolution of prefrontal cortex. *Neuropsychopharmacology, 47*(1), 3–19.

Rosa, M. G., & Tweedale, R. (2005). Brain maps, great and small: Lessons from comparative studies of primate visual cortical organization. *Philosophical Transactions of the Royal Society London B: Biological Sciences, 360*(1456), 665–91.

Schimmack, U., Heene, M., & Kesavan, K. (2017, February 2). Reconstruction of a train wreck: How priming research went off the rails. https://replicationindex.com/2017/02/02/reconstruction-of-a-train-wreck-how-priming-research-went-of-the-rails/.

Seth, A. (2021). *Being you: A new science of consciousness.* London: Faber and Faber.

Simons, D. J., & Chabris, C. F. (1999). Gorillas in our midst: Sustained inattentional blindness for dynamic events. *Perception, 28,* 1059–74.

## Chapter 12—Free Will

Anderson, N. E., & Kiehl, K. A. (2020). Re-wiring guilt: How advancing neuroscience encourages strategic interventions over retributive justice. *Frontiers in Psychology, 13*(11), 390.

Aono, D., Yaffe, G., & Kober, H. (2019). Neuroscientific evidence in the courtroom: A review. *Cognitive Research: Principles and Implications, 4*(1), 40.

Bergson, H., & Pogson, F. L. (1959). *Time and free will: An essay on the immediate data of consciousness.* London: George Allen and Unwin.

Button, K. S., Ioannidis, J. P., Mokrysz, C., Nosek, B. A., Flint, J., Robinson, E. S., & Munafò, M. R. (2015). Power failure: Why small sample size undermines the reliability of neuroscience. *Nature Reviews Neuroscience, 14*(5), 365–76.

Churchland, P. S. (2011). *Braintrust: What neuroscience tells us about morality.* Princeton, NJ: Princeton University Press.

Dennett, D. C. (2017). *From bacteria to Bach and back: The evolution of minds.* New York: W. W. Norton.

Dennett, D. C., & Caruso, G. D. (2021). *Just deserts: Debating free will.* Cambridge: Polity.

Dretske, F. (1988). *Explaining behavior: Reasons in a world of causes.* Cambridge, MA: MIT Press.

Duncan, L. E., & Keller, M. C. (2011). A critical review of the first 10 years of candidate gene-by-environment interaction research in psychiatry. *American Journal of Psychiatry, 168,* 1041–49.

Ellis, G. F. R. (2016). *How can physics underlie the mind? Top-down causation in the human context.* Berlin: Springer.

Farahany, Nita A. (2016). Neuroscience and behavioral genetics in US criminal law: An empirical analysis. *Journal of Law and the Biosciences,* 2(3), 485–509.

Flint, J., & Munafò, M. R. (2013). Candidate and non-candidate genes in behavior genetics. *Current Opinion in Neurobiology, 23,* 57–61.

Frith, U., and Frith, C. (in press). *What makes us social.* Cambridge, MA: MIT Press.

Glenn, A. L., & Raine, A. (2014). Neurocriminology: Implications for the punishment, prediction and prevention of criminal behaviour. *Nature Reviews Neuroscience, 15*(1), 54–63.

Heyes, C. M., Bang, D., Shea, N., Frith, C. D., & Fleming, S. M. (2020). Knowing ourselves together: The cultural origins of metacognition. *Trends in Cognitive Sciences, 24,* 349–62.

Kim, J. (1993). *Supervenience and mind: Selected philosophical essays.* Cambridge: Cambridge University Press.

Kim, J. (1998). *Mind in a physical world: An essay on the mind–body problem and mental causation.* Cambridge, MA: MIT Press.

Laland, K. (2017). *Darwin's unfinished symphony: How culture made the human mind.* Princeton, NJ: Princeton University Press.

Meincke, A. S. (2018). Autopoiesis, biological autonomy and the process view of life. *European Journal for Philosophy of Science, 9,* 5.

Mitchell, K. J. (2018). Does neuroscience leave room for free will? *Trends in Neuroscience, 41*(9), 573–76.

Nahmias, E. (2014). Is free will an illusion? Confronting challenges from the modern mind sciences. In Walter Sinnott-Armstrong (Ed.), *Moral psychology: Vol. 4. Free will and moral responsibility.* Cambridge, MA: MIT Press.

Nicholson, D. J., & Dupré, J. (Eds.). (2018). *Everything flows—Towards a processual philosophy of biology.* Oxford: Oxford University Press.

Pickering, A. (2010). *The cybernetic brain: Sketches of another future.* Chicago: University of Chicago Press.

Potter, H. D., & Mitchell, K. J. (2022). Naturalising agent causation. *Entropy,* 24(4), 472.

Rovelli, C. (2021). *Helgoland*. London: Allen Lane.
Scurich, N., & Appelbaum, P. S. (2017). Behavioral genetics in criminal court. *Nature Human Behavior, 1*, 772–74.
Steward, H. (2012). *A metaphysics for freedom*. Oxford: Oxford University Press.
Tse, P. U. (2013). *The neural basis of free will: Criterial causation*. Cambridge, MA: MIT Press.
Von Bertalanffy, L. (1969). *General system theory: Foundations, development, applications*. New York: G. Braziller.
Whitehead, A. N. (1978). *Process and reality*. New York: Free Press.
Wiener, N. (1961). *Cybernetics or control and communication in the animal and the machine*. Cambridge, MA: MIT Press.
Yarkoni, T. (2009). Big correlations in little studies: Inflated fMRI correlations reflect low statistical power. Commentary on Vul et al. (2009). *Perspectives in Psychological Science, 4*(3), 294–98.

## Epilogue—Artificial Agents

Coelho Mollo, D. (2022). Intelligent behaviour. *Erkenntnis*, 1–17.
Eliasmith, C. (2013). *How to build a brain: A neural architecture for biological cognition*. Oxford: Oxford University Press.
Farnsworth, K. D. (2017). Can a robot have free will? *Entropy, 19*(5), 237.
Froese, T., & Taguchi, S. (2019). The problem of meaning in AI and robotics: Still with us after all these years. *Philosophies, 4*(2), 14.
Garnelo, M., & Shanahan, M. (2019). Reconciling deep learning with symbolic artificial intelligence: Representing objects and relations. *Current Opinion in Behavioral Sciences, 29*, 17–23.
Gopnik, A. (2017). Making AI more human. *Scientific American, 316*(6), 60–65.
Jonas, H. (1968). Biological foundations of individuality. *International Philosophy Quarterly*, 8, 231–51.
LeCun, Y., Bengio, Y., & Hinton, G. (2015). Deep learning. *Nature, 521*(7553), 436–44.
Legg, S., & Hutter, M. (2007). A collection of definitions of intelligence. *Frontiers in Artificial Intelligence and Applications, 157*, 17.
Marcus, G. (2018). Deep learning: A critical appraisal. *arXiv*, 1801.00631.
Marcus, G. (2022). Noam Chomsky and GPT-3: Are large language models a good model of *human* language? https://garymarcus.substack.com/p/noam-chomsky-and-gpt-3.
McCulloch, W. S., & Pitts, W. (1943). A logical calculus of the ideas immanent in nervous activity. *Bulletin of Mathematical Biophysics, 5*, 115–33.
Mitchell, M. (2019). *Artificial intelligence: A guide for thinking humans*. New York: Farrar, Straus, and Giroux.
Pearl, J. (2018). Theoretical impediments to machine learning with seven sparks from the causal revolution. *arXiv*, 1801.04016.
Sejnowski, T. J. (2020). The unreasonable effectiveness of deep learning in artificial intelligence. *Proceedings of the National Academies of Science USA, 117*(48), 30033–38.
Zador, A., Richards, B., Ölveczky, B., Escola, S., Bengio, Y., Boahen, K., Botvinick, M., et al. (2022). Toward next-generation artificial intelligence: Catalyzing the NeuroAI revolution. *arXiv*, 2210.08340.

# INDEX

Note: page numbers in italics refer to figures.